AGRICULTURAL TERMINOLOGY

About the Authors

A.K. Singh, is Assistant Professor (Agriculture) at North Eastern Regional Institute of Water and Land Management, Tezpur , Assam. He did his M. Sc. from Birsa Agricultural University, Ranchi, and Ph. D. Visva Bharati Santiniketan. He is a recipient of University Fellowship, Department of Atomic Energy Fellowship and National Agricultural Technology Project Fellowship. He has published 16 research papers in National and International journals of repute and attended several seminars.

N.P. Singh, is presently working as a Principal Scientist and Head, Division of Agronomy, ICAR Research Complex for NEH Region, Umiam, Meghalaya. Earlier he was Professor and Head, Department of Agronomy, School of Agricultural Science and Rural Development, Nagaland University, Medziphema. He completed his M.Sc. and Ph. D. in Agronomy from the Institute of Agricultural Sciences, Banaras Hindu University, Varanasi. He is a recipient of District Merit Fellowship and BHU Fellowship. He has published 38 research papers and 70 popular articles. He has also attended 76 conferences at regional, national and international levels.

AGRICULTURAL TERMINOLOGY

A.K. SINGH

N.P. SINGH

CONCEPT PUBLISHING COMPANY, NEW DELHI-110059

ISBN : 81-8069-066-0 (HB)

First Published 2004

Published and Printed by
Ashok Kumar Mittal
Concept Publishing Company
A/15-16, Commercial Block, Mohan Garden
New Delhi-110059 (INDIA)

Phone: 25351460, 25351794
Fax: 091-11-25357103
Email: publishing@conceptpub.com

Foreword

This book is a compilation of terminology used in the field of agricultural research, education and extension. The terminologies included in this handbook are of varying degree and with varying meanings by teachers, students, researchers, extension workers, planners and others in agricultural development.

I am pleased to have the honour of commenting on this important publication. This handbook provides a convenient and concise source of precise definitions of all possible common terms used in agriculture under one cover. Efforts have been made to compile recent available information so as to make it more useful. Undoubtedly, it will need to be revised as more new terms come into use and meanings of existing terms are modified. Meanwhile, it should find a place on our desk and in libraries.

Shri P.L. Thanga
Secretary,
North Eastern Council,
Shillong

Preface

This handbook offers a comprehensive coverage on important agricultural terms that is useful to students in curricula, such as agronomy, plant breeding, horticulture, soil science as well as to non-agricultural students of plant science, ecology, natural resources and environmental studies.

It is a common feature that the same terminology is defined differently by different authors and thus, the best possible definitions were selected or developed and updated. While compiling the authors have drawn some matter from popular journals, books, report, dictionaries/glossaries of terminology, etc., and is indebted to the editors, publishers and authors of these publications.

Efforts taken for preparation of this publication by Ms. Reena Singha, Ms. Rupali Hazarika and Sri Iswar Das are appreciated.

We would be highly grateful if readers would draw our attention for further improvement of this publication.

May, 2003

A.K. Singh
N.P. Singh

A

Abiotic factors: The non-living components of the environment that directly affect plant and animal life, such as water, carbon dioxide and light.

Aboricide: A chemical used for killing trees or woody shrubs.

Abscission: The shedding of fruits, leaves or stems from the parent plant.

Absolute humidity: Grams of water vapour in the air per cubic metre.

Absolute transpiration: The rate of water loss from a plant determined experimentally.

Absorption active: Movement of ions and water into the plant root as a result of metabolic processes by the root, frequently against an electro-chemical potential gradient.

Absorption passive: Movement of water into roots resulting from pulling forces on the water column in the plant as water is lost to the atmosphere through the leaves by transpiration.

Abstract conceptualization: The stage of learning cycle where people use logical analysis of ideas and rely on systematic planning to build learning experience.

Acclimatization: An adjustment in the morphology or physiology of an organism in response to a change in the environment.

Accommodators: Trainees with accommodative learning style and rely on concrete experience and active experimentation.

Accumulation: The uptake of substance against a concentration gradient and hence an energy requiring process.

Acedophiles: The plants which can grow in acidic soils (*p*H range 4.5–6.5).

Acentric Chromosome: A chromosome fragment without centromere.

Achene: A dry, hard, one-seeded indehiscent fruit in which the seed is attached to the ovary wall at a point.

Acid: A substance that forms hydrogen ions (H^+) when dissolved in water (a proton donor).

Acid equivalent (a.e.): refers to that part of a formulation that theoretically can be converted into the acid. The theoretical yield of acid from an active ingredient.

Acid forming fertilizer: A term applied to commercial fertilizers which leave an acid residue in the soil. The amount of calcium carbonate required to neutralize the acid residue is referred as its equivalent acidity.

Acid salts: When only a base replaces a part of the replaceable hydrogen in an acid, an acid salt is obtained. The acid salt possesses the properties of salt in addition to those of acids. They react with alkalis to give normal salt and water.

Acid soil: A soil with *p*H of less than 7.0 (usually less than 6.6). An acid has a preponderance of hydrogen ions over hydroxyl ions.

Acidic cations: Hydrogen ions or cations that, on being added to water undergo hydrolysis resulting in an acidic solution. Examples in soils are H^+, Al^{3+} and Fe^{3+}.

Acidulation: The process of treating a fertilizer source with an acid. The most common process is treatment of phosphate rock with an acid such as sulphuric, nitric, or phosphoric acid.

Acquired character: A feature induced during the life of an individual by environmental factors.

Acre: An area of 43.560 ft^2 or 0.405 ha.

Acre-foot of water: It is the amount of water that would cover an acre of land to a depth of one foot, assuming no seepage, evaporation and run off losses.

Acre-inch: It is a measure of quantity of flow of water and is equal to flow which will cover one acre to a depth of one inch.

Acrocentric Chromosome: A chromosome with its centromere near one end.

Acropetal: Towards the apex of a plant organ; generally upwards in shoots and downwards in roots. Opposite of basipetal.

Actinomycetes: A non-taxonomic term applied to a group of organisms with characteristics intermediate between the simple bacteria and the true fungi.

Activation energy: The energy required before the reaction of concern will proceed. Often, heating will supply the needed energy. Enzymes or other catalysts lower the activation energy of a reaction.

Activator: A substance that accelerates the effect or increases the total effect of a product.

Active absorption: Absorption of water and other substances against the concentration gradient and therefore, involving expenditure of energy into inhibition and osmosis.

Active acidity: The activity of hydrogen ions in the aqueous phase of soil. Its activity is measured and expressed in *p*H value.

Active experimentation: The stage of learning cycle where learning takes place through experimenting with new environments.

Active ingredient: It is that part of a chemical formulation which is directly responsible for the herbicidal effects. In some herbicides the entire molecule is considered as the active unit. Thus, if the chemical is 95 per cent pure it will be considered as having 95 per cent active ingredient. In others, the herbicidal activity is more accurately calculated on an acid equivalent basis.

Actual farm yield: The yield of a crop/variety on the farmer's field under optimum management practices.

Actual vapour pressure: Pressure exerted by water vapour contained in the air; unit millibar (mb) or mm of Hg.

Adaptability: The capacity of adjusting a farm system to cope with changing conditions.

Adaptation: It is the process by which organisms become more suited to survive and function in a given environment.

Addition line: An addition line has one pair of chromosomes from another variety or species in addition to the normal somatic chromosome complement (2n) of the species.

Additive: Substance intended to improve the properties of a fertilizer or soil conditioner.

Additive Gene Effect: In case of polygenes, the effects of different genes affecting a character being cumulative or additive in nature.

Adhesion: The force of attraction that binds the molecules of different kinds. Molecular attraction holds the surface of two dissimilar substances in contact such as water and rock particles.

Adjuvant: An additive ingredient that facilitates or modifies the action of the principle ingredient.

ADP (adenosinediphosphate): Adenosine-derived ester formed in plant cells and converted to ATP for energy storage.

Adsalt sugarcane: Sugarcane that takes 18 months for harvesting usually planted in July.

Adsorbed: Held so tightly that the herbicide is either inactive or very slowly effective.

Adsorption: The process by which atoms, molecules, or ions are taken up and retained on the surfaces of solids by chemical or physical binding (*e.g.,* the adsorption of cations by negatively charged minerals).

Advance phase (irrigation): The portion of the total irrigation time during which water advances in over land flow surface from the upper end of the field to the lower end of the field. The time elapsed during this phase is called advance time.

Advection: The process of transfer (of an air-mass property) by virtue of motion.

Advection fog: Fog formed by passage of relatively warm, moist and stable air over a cool surface.

Adventitious embryony: A type of apomixis where the embryos arise from a cell or a group of cells either in the nucellus or in the integument outside the embryo sac in addition to the regular sexual embryo.

Aerate: To allow or promote exchange of soil gases with atmospheric gases.

Aeration: Process of ventilating soil or crops to attain a uniform temperature that is near the ambient air temperature and encourages the development of soil microorganisms.

Aeration (soil): The process by which air in the soil is replaced by air from atmosphere. In a well-aerated soil the soil air is similar in composition to atmosphere above the soil. Poorly aerated soils usually contain a much higher percentage of carbon dioxide and a corresponding lower percentage of oxygen than the atmosphere above the soil. The rate of aeration depends largely on the size, volume, and continuity of pores not filled with water from the surface and within the soil.

Aeration (storage): The process of movement of air through grain in storage at low air flow rate for the primary purpose of maintaining or improving the quality of the grain.

Aeration porosity: The proportion of the bulk volume of the soil that is filled with air when the moisture tension is at some specified value. The tension is usually specified in the range of 40 to 100 cm of water.

Aerial map: Aerial map offers a prospective view of portions of the land surfaces, they do not have a true scale. These maps cover large areas and are taken either from an aircraft or a satellite. They are used to interpret soil boundaries, land forms, soil characterising factors like vegetation, soil colour, soil types, drainage and slopes. Aerial maps are available with the National Remote Sensing Agencies.

Aerobic: (*i*) Having molecular oxygen as a part of the environment; (*ii*) Growing only in the presence of molecular oxygen, as aerobic organisms; (*iii*) Occurring only in the presence of molecular oxygen.

Aeroculture: Cultivation of woody plants particularly those used for decoration and shade.

Aerodynamic: Refers to forces of moving air acting upon the soil or liquid or crop surface.

Aeroponics: It is a system for growing plants with their roots supplied with moisture in the air. Plants like Citrus and Olive have been successfully grown through this method.

Aerosol: An aggregate of minute particles (solid or liquid) suspended in the atmosphere.

Aerosols: Small particles of herbicides or a liquid carried as a dispersion in a gaseous phase.

Affective learning: Learning that does with the way we feel about tasks.

Afforestation: Artificial establishment of a forest on land which had not previously grown tree crops on deforested land.

Aflatoxins: Potent carcinogens in nature that are antibodies produced by fungal growth on many grains and nuts.

After cultivation: Harrowing, rolling, tilling and other cultivation carried out in a field after the crop has emerged.

After ripening: A term for the collective physiology changes that occur in a dormant seed which makes it capable of germination.

Aftermath: The second growth of meadow plants after the first crop has been cut.

Agglutinin: An antibody present in blood serum specific to an antigen present on red blood cells (RBC); it causes the agglutination of RBCs having the concerned antigen. (*Syn.*, antibody).

Agglutinogen: An antigen present on RBCs; it reacts with specific agglutinins to cause an agglutination of RBCs. (*Syn.*, antigen).

Aggregate: A unit of soil structure, usually formed by natural processes in contrast with artificial processes, and generally <10 mm in diameter.

Aggregate fruit: Made up of two or more carpel from a single flower and the stem axis; e.g., blackberry, strawberry.

Agricultural drought: It occurs where soil moisture and rainfall are inadequate during the growing season to support healthy crop growth to maturity and cause extreme crop stress and wilt.

Agricultural forest: These are mixtures of tree crops, meadows domestic animals living fences (hedge and wind breaks), other perennial plants that produce food, fibre and chemicals. Thus forests are designed to simulate the basic processes of a natural forest ecosystem for yielding human foods and materials.

Agricultural statistics: It concerns with scientific methods for collecting, organizing, summarizing, presenting and analysing data generated from agricultural experiments as well drawing valid conclusion and making reasonable decision on the basis of such analysis.

Agricultural waste: The characteristics of the agricultural waste materials make them important in respect of their use and economic value. Agricultural wastes can be put to use because: (*i*) it is the source of energy and power; (*ii*) it acts as animal feed and fodder; (*iii*) it is the human food; and (*iv*) it can be used for farm development (improvement of soil health, soil fertility, soil physical conditions and plant protection).

Agriculture: An activity of man, primarily aimed at the production of food, fibre, fuel etc. by optimum use of terrestrial resources.

Agrihorticulture: Agrihorticulture is a form of agroforestry where fruit trees are the tree components. This system provides for a higher income per unit area, with the additional income being generated from fruit yield. It also serves as an option to the farmer even in case of crop failure. Sometimes, well maintained and well established orchard bring even higher returns than the field crops. Fruit trees also provide protection to degraded lands. The species of fruits to be grown depend largely on the amount of rainfall. Porous, aerated and deep soils are good for fruit growing. These systems require high initial investment and care.

Agrobiology: A phase of the study of agronomy dealing with relation of yield to the quantity of an added or available fertilizer element.

Agroclimatic region: The grouping of different physical areas within the country into broadly homogenous zones based on climatic and edaphic factor.

Agro-Climatic Zones: Refer to "Agro-Ecological Regions of India" (NBSS & LUP Publication and publications of ICAR) on agro-climatic classification for NARP zones and Planning Commission classification of agro climatic zones.

Agroecological zone: A major area of land that is broadly homogenous in climatic and edaphic factors, but not necessarily continuous, where a specific crop exhibits roughly the same biological expression.

Agroecology: The holistic study of agroecosystem, including all environmental agricultural crops and human elements, their interrelationships and the process in which they are involved.

Agroecosystem: It is composed of total complex of the crop/animal in area together with overall environment and modified by management practices to produce food, fibre, fuel and other products desired for human use.

Agroenergy: It is a broad term used for referring to energy issues related to agriculture, both as a producer of energy as well as a user of energy, is captured and stored as biomass through the process of photosynthesis by plants which could be used as a direct source of energy in the form of food, feed and fuel and could be converted to gaseous or liquid forms of fuels for efficient use.

Agro-forestry: (*i*) Agro-forestry is a sustainable land development system which combines the cultivation of trees along with crops. It can be adopted in all the agroclimatic zones. The species of trees planted depend on the requirements of the user for fuel, fodder, timber or food. Inclusion of trees within an agricultural land use system yield both tree and food products thus meeting the various needs of the user. It also restores the productivity of the land use by increasing the inflow of water into the soil.

(*ii*) The deliberate use of woody perennials (trees, shrubs, palms, bamboo) on the same land-management unit as arable crops, pastures and/or animals, either in a mixed spatial arrangement in the same place at the time, or in a sequence over time. A land-use to benefit from resultant ecological and economic interactions.

Agrology: The study of applied soil science and soil management.

Agrometeorology: The branch of meteorology relevant to problems of agriculture. It is concerned, in particular, with the surface layer of the atmosphere and with condition with the top layer of the earth's surface which are associated with the variations of the meteorological element like rainfall, humidity, temperature, sunshine, wind velocity, atmospheric pressure etc.

Agronomic trials: Evaluation of new strains for release as a variety; to determine the optimum date of sowing, fertilizer level, irrigation level etc.

Agronomy: The word has been derived from the Greek word 'agros' means fields and 'nomos' means management. It is a specialized branch of agriculture dealing with crop production and soil management.

Agropisciculture: Combining cropping and the controlled breeding, hatching and rearing of fish within a farm.

Agrosilvipastoralism: Land use system in which arable cropping, use of woody vegetation and the keeping of grazing livestock are combined.

Agrostology: It is the science of a mineral soil having maximum biological activity or eluviation (removal of materials dissolved or suspended in water or both).

Air capacity of soil: The quantity of air present in the soil at field capacity.

Air dry: (*i*) The state of dryness at equilibrium with the water contents in surrounding atmosphere. The actual water content will depend upon the relative humidity and temperature of the surrounding atmosphere; (*ii*) To allow to reach equilibrium in water content with the surrounding atmosphere.

Air layering: A method of layering which involves the rooting of an attached stem by excluding light, under moist conditions, from a portion of the stem.

Air mass: A body of air in which horizontal gradient of temperature and humidity are relatively high and separated from adjacent body of air by a more or less sharply defined transition zone.

Air porosity: The proportion of the bulk volume of soil that is filled with air at any given time or under a given condition, such as a specified moisture tension.

Air-borne diseases: These diseases usually affect growing plant. Treatment by applying chemicals to foliage is the most common way of controlling air-borne bacterial and fungal diseases. Viral diseases are yet to be fully controlled by this method. Control measures for the various diseases include *Cultural control* like crop rotation, mixed cropping, adjustment of sowing depth and time. Sanitizing the field by removing weeds and diseased debris. *Biological control* such as use of alternate and collateral hosts. *Chemical control* includes the use of fungicides and other chemical treatments.

Aitonomic (stimulative) parthenocarpy: A type of parthenocarpy in which fruit development involves pollination but no fertilization. The stimulus of pollen is required to develop the ovary into fruit.

Albedo: The ratio of the amount of solar radiation reflected by a body to the amount incident upon it, often expressed as a percentage, as the albedo of the earth is 34 per cent.

Albinism: The absence of melanin pigment in skin, hairs, eyes etc. of an animal; the absence of chlorophyll in a green plant.

Alder: *Alnus nepalensis,* a tree species endemic to the sub-Himalayan region and well sought for its nitrogen-fixing capabilities, timber quality and ability to coppice.

Alder-based system: The system of cultivating crop with alder trees.

Algae: A heterogeneous group of photosynthetic unicellular and multicellular organisms comes under *Thallophyta.*

Aliphatics: The compounds derived from straight chain hydrocarbons.

Alkaline (or basic) fertilizer: A fertilizer capable of increasing the residual alkalinity by reducing the residual acidity of the soil.

Alkaline fan: Fan-shaped alluvium deposited at the mouth of ravines or canyons as the slope lessens and water-carried debris settles out of the slowed water.

Alkaline forming: A term applied to commercial fertilizer that leaves alkaline or basic residues in the soil. The basic residual effect is expressed in terms of 'equivalent basicity'. More the basic residue, more will be the equivalent basicity.

Alkaline soil: A soil that is alkaline in reaction throughout the root zone or for a major part of the root zone. Precisely, any soil having a *p*H value greater than 7.0. Practically, a soil having a *p*H above 7.3 is called alkaline.

Allele: An alternative form of a gene; alleles occupy the same position or locus on the homologous chromosomes; usually, alleles do not show complementation and yield typical monohybrid ratio in F_2.

Allelopathy: Direct/indirect harmful effect that one plant has on another or manually on each other through the production of chemical compounds that escape into the environment.

Alley cropping: A farming system in which arable crops are grown in alley formed by trees or shrubs, established mainly to hasten soil-fertility reiteration and enhance soil productivity. Annual crops are grown in spaces between rows and trees or shrubs, often leguminous one that tolerate heavy and regular coppicing. The leafy and woody material of the trees and shrubs is used as mulch in the crop(s) and also often as fodder, timber, fuel etc. Also referred to as venue cropping and hedgerow cropping.

Allocation games: Games related to resource allocation problems where the resources like money, power etc. are scarce.

Allogamy: In allogamy, pollen grains from flowers of one plant pollinate the flowers of other plants (*Syn.*, cross-pollination).

Allopolyploid: A polyploid having two or more distinct genomes; usually produced by chromosome doubling of interspecific hybrids (F_1S).

Allosteric effect: The interaction of a small molecule with a protein molecule at one site modifies the activity of another site(s) of the same protein molecule.

Allotetraploid: An allopolyploid having four genomes; usually two copies each of two genomes; may be four different genomes.

Alluvial soil: A soil developed from recently deposited by alluvium (running water of streams and rivers) and exhibiting essentially no horizon development or modification of the recently deposited materials. This is the most important soil group in India, derived from the deposition from tributaries of the Ganga, Brahmaputra and Indus. Despite a great deal of variation, alluvial soils, primarily include deltaic alluvium, calcareous alluvial soils, coastal alluvium and coastal sands. Geologically the alluvium is divided into *khadar* or the newer alluvium which is sandy, generally light coloured and less stony (*kankari*) in composition, and *bhangar* or the older alluvium which is dark, more clayey and full of stones (*kankar*). The soils range from drift sand to loams and from fine silts to stiff clays. Occasional pebbles are also present. The drawbacks of these soils are—presence of impervious clays in part obstructs drainage and promotes accumulation of injurious salts of sodium and magnesium making these soils less fertile. Formation of hard pan at certain levels in the soil profile through binding of soil grains by alluviation of finer soil is sometimes observed.

Alpha decay: The radioactive transmutation of an element into another by the emission of an alpha particle.

Alpha radiation: The radiation consisting of alpha particles emitted by certain radioactive elements.

Alpha ray: A ray of positively charged particles emitted during certain radioactive transformations.

Alternative drying and wetting (rice): Frequent draining and reflooding (aerobic and anaerobic cycle) in rice fields.

Altitude: Vertical distance of a level, a point or an object considered as a point measured from the mean sea level.

Aluminium toxicity: It is caused by excess water soluble and exchangeable aluminium, usually on acid sulphate soils and strongly acidic; soil symptoms are white or yellow interveinal blotches on the leaves, the leaves dry out and die, roots are short and scanty and plants are stunted.

Aman rice: The word 'aman' is derived from Arabic, means 'safety' which indicate stability of the crop. Aman rice is direct seeded in pre-monsoon period (March–April) in the deep-water (1-6 m deep) areas. The rice plants grow with rising floodwater and settles at the ground where water subsidies during the post-monsoon period. This rice crop is photosensitive and is harvested during October–December.

Ambient temperature: The temperature of surrounding atmospheric air.

Ambimobile: The character of chemical substances (*e.g.*, herbicides) which, when absorbed by the plant, move in both the phloem and xylem.

Ameliorate: To make better, especially more tolerant.

Amendment, soil: Any material such as lime, gypsum, sawdust, sulphur or soil conditioner, that is added to the soil to influence plant growth by changing the soil reaction or *p*H or by improving the physical condition of the soil. Strictly, a fertilizer is also an amendment, but the term is used most commonly for added materials other than fertilizers.

Amensalism: Production by one organism of a substance that is inhibitory to other organisms. Similar in use to the term *allelopathy.*

Amines: A class of compounds derived from ammonia by replacing the hydrogens with organic radicals.

Amino acids: These are building blocks of protein and each amino group ($-NH_2$) and at least one carboxyl ($-COOH$) molecule.

Ammonia (liquid/anhydrous): A material mostly produced by the synthetic process and obtained in the liquid form. It is used in the fertilizer industry for ammoniation of superphosphate, in making base mixtures, in making mixed fertilizers, or for conversion into salts such as sulphate, phosphate, chloride, nitrate etc. It is also applied directly into the soil by suitable mechanical means. Fertilizer grade anhydrous ammonia contains about 82 per cent of nitrogen.

Ammonia volatilization (soil): Ammonia readily identifiable product of nitrogen mineralisation, formed continuously in the soil and water of a flooded paddy. Organic matter decomposition in absence of oxygen and high *p*H favours ammonia volatilization from flooded soil.

Ammoniated superphosphate: A product obtained when superphosphate is treated with ammonia or with solution which contains ammonia or other compounds of nitrogen.

Ammoniating solution (nitrogen solution): A solution used for ammoniating superphosphate or a mixture of superphosphate with other fertilizers. This solution may be liquor ammonia itself or a solution of ammonia nitrate or urea in liquor ammonia.

Ammonification: Production of ammonia as a result of biological decomposition of organic nitrogen compounds.

Ammonium diffusion (soil): Diffusion of NH^{+4} from the anaerobic soil layer to the aerobic soil layer where it undergoes nitrification and denitrification.

Ammonium fixation: Absorption of ammonium ions by soil or minerals in such form that they are neither water-soluble nor readily exchangeable.

Ammonium nitrate (NH_4NO_3): The ammonium salt of nitric acid containing 33 per cent nitrogen, half of which is the ammonium form and the other in the nitrate form. It is in the granular form or pills form.

Amorphous clays: Lacking adequate crystal orientation to diffract X-rays. Without regular crystalline form in molecular size.

Amorphous material: Non-crystalline soil materials.

Amphidiploid: An allopolyploid having two copies each of two or more distinct genomes; thus an amphidiploid has the somatic chromosome complement of two or more diploid species, natural amphidiploids behave as diploids during meiosis, *i.e.*, they show normal bivalent formation, *e.g.*, wheat, tobacco, oats, cotton etc.

Anabolism: An energy consuming process, wherein complex organic substances are produced from simple materials.

Anaerobe: An organism that survives in the absence of air or molecular oxygen.

Anaerobic: The absence of molecular oxygen.

Analysis of variance (Anova): This is a statistical technique by which source of variation of any attribute may be partitioned and estimated.

Anaphase: The stage of cell division following metaphase; homologous chromosomes (A I of meiosis) or sister chromatids (A II of meiosis and in mitosis) begin to move from the equatorial plate towards the opposite poles.

Andragogy: The study of adult learning.

Androdioecious: A sex form in dioecious species where only staminate flowers are borne on one plant and bisexual flowers on another plant of the same variety or species.

Androecious (staminate): A sex form where only male flowers are produced, giving rise super male plant.

Androecium: The stamens of a flower collectively.

Andromonecious: A sex form where staminate and hermaphrodite flowers are separately produced in the same plant.

Anemometer: Instrument used in the measurement of the wind speed.

Aneroid barometer: An instrument, not using any liquid, used for measuring atmospheric pressure. Its advantage over a mercury barometer is its portability.

Aneuploid: An organism or a cell having one or few chromosomes more or less than the normal disomic number of the species; the chromosome number is not a multiple of the basic number (*x*).

Angiosperm: One of the flowering plants whose seeds are enclosed in fruits.

Anhydrous: Dry or without water.

Animal manure: There are several products of animal origin that have manure value: (*a*) Fish manure, bird guano, and animal dung are used for vegetable and horticultural plants. (*b*) Activated sludge is a product wherein the liquid wastes of cities are processed in treatment plants. (*c*) Sheep and goat droppings are also used as manure.

Anion exclusion: The exclusion or repulsion of anions from the vicinity of soil particle surfaces because of the negative potential associated with most soil particles.

Anionic surfactant: It is one that has negative charge and performs best in soft water. Most wetting agents and detergents and emulsifiers are anionic.

Annual: A plant completing its life cycle in one season or one year.

Anomaly: The departure of an element from its long period average value for the place concerned.

Antagonism: The action of two or more chemicals in a mixture, with a total effect smaller than the most active component applied alone.

Anther culture: Culture of anthers (or pollen grains) on a suitable medium for the production of callus and/or haploid plants.

Anthesis: The first opening of a flower.

Anthocyanin: Any of a class of water-soluble pigments, including most of those imparting red or blue colour to fruit or flowers.

Antibiosis: The adverse effect of feeding on a resistant host plant on the development and reproduction of an insect pest.

Antibiotic: An organic compound produced by micro-organisms or higher plants which in sufficient concentration will kill or inhibit growth of certain other organisms.

Antibody: A protein molecule that interacts specifically with an antigen molecule; usually produced by an animal in response to the antigen in question.

Anticodon: The three bases of a tRNA that are complementary to the three bases of mRNA coding for an amono acid.

Anticyclone: The atmospheric pressure distribution in which there is a high central pressure relative to the surroundings.

Antigen: A substance, usually a protein, which induces specific antibody production when introduced into a living system, usually an animal, in which it does not occur naturally.

Antimetabolite: A chemical, which interferes with the metabolic activities of an organism, *e.g.*, antibiotics.

Antitranspirants: The chemicals which either partially close the stomata or form a thin film over the produce and check water losses through transpiration.

Apex: Top or highest point.

Apical dominance: The apical dominance is the phenomenon in the plant where the apical shoots/buds have inhibitory influence on the growth of lateral shoot and buds.

Apiculture: The art of raising bees for the production of honey on a commercial scale is called apiculture. The bees are raised in scientifically fabricated wooden boxes which serve as artificial beehives. The place where the bees are kept in wooden boxes is known as an apiary. The person looking after the activity is known as an apiarist.

Apogamy: Development of embryo from synergids or antipodal cells without fertilization; a form of apomixis.

Apomixis: Development of embryo (and seed) without fertilization.

Apoplast: The continuous, non-living cell wall structure that surrounds the symplast, forming a continuous translocation system.

Apospory: A form of apomixis in which the embryo sac develops from a vegetative cell of the ovule.

Appendage: An additional part of a structure of organ.

Apple scald: Accumulation of toxic levels of volatile in the skin tissues leading to superficial death and burrowing of affected tissues; a storage disorder where immature fruit and the greater side of the fruit is most affected.

Aquaculture: The deliberate production for human use plant and animal organisms living in water.

Aquatic: A plant that grows in water. There are three kinds: *Sumergent* grows beneath the surface such as water weed and green algae; *emergent* grows above the water, such as cattails and water lilies; *floaters* such as water hyacinth and water lettuce.

Aquatic weeds: Weeds which are found in lakes and ponds.

Aquic: A soil water regime, mostly too wet (reducing conditions, waterlooged) for parts of the year.

Aquifer: It is the geological or oil, layer formation on having ground water.

Arable crops: Those crops which require cultivation.

Arable farming: Farming system that involves the production of crops requiring tillage.

Arable land: Lands that are suitable for the cultivation and production of food crops. They include 4 classes: *Class 1-Characteristics* (level land, depth above 90 cms, well drained, easily workable soils, good water holding capacity, highly fertile, suitable for intensive cultivation, requires no special care; *Class 2-Characteristics* (gentle slopes, moderate erosion, moderately fertile soils, soil depth: 22.5 cm to 45 cm, imperfect for rapid drainage, soils difficult to work with, due to the presence of stones, rocks and pans, cultivation can be practised with soil conservation measures). *Class 3-Characteristics* (moderately steep slopes, very severe erosion rate, low moisture retentivity, frequent or long duration overflow, shallow soils, very heavy or light, farming operations difficult due to the presence of a hard pan, severe limitations for permanent cultivation); *Class 4-Characteristics* (suitable only for occasional cultivation, best maintained by perennial vegetation because of very unfavourable characteristics like slope, erosion, depth and drainage, maintenance requires very costly measures).

Areole (islet): A small area of mesophyll delimited by intersecting veins.

Arid: A term applied to regions or climates that lack sufficient moisture for crop production without irrigation. A climate in which potential evaporation exceed rainfall in all months of the year so that cropping is possible only with the support of water harvesting or irrigation; refers to an area with an average of less than about 200 mm annual rainfall.

Aridic: A soil water regime with long dry periods.

Aril: An additional covering formed on certain seeds by expansion of the stalk of the ovule after fertilization.

Aromatic: Chemical compound derived from the hydrocarbon benzene (C_6H_6) and having a pungent odour.

Artificial fertilizers: Fertilizers made by chemical process or mined. Also known as inorganic/mineral fertilizers/chemical fertilizers.

Artificial selection: Selection by man.

Asexual reproduction: (*i*) Reproduction of an organism without involving the germ or sexual cells, such as by astolons (vegetative) or apomaxic seeds in certain plants.

(*ii*) The progeny are produced through a process which does not involve the fusion between male and female gametes.

Ash: The mineral residues remaining after the destruction of organic material by burning.

ASL: Above Sea Level.

Assimilation: The processes where by photosynthesis is utilized in the building up of protoplasm and cell wall.

Assimilators: Trainees with assimilative skill rely more on abstract conceptualization and reflective observation.

Astringent: Substance that binds or draws together soft tissues.

Asymmetrical hybrids: Cells/plants produced through somatic hybridization and containing the full somatic complement of one fusion partner but only a part of the somatic complement of the other.

Asynapsis: The failure, partial or complete, of pairing (synapsis) between homologous chromosomes during meiosis (zygotene of prophase I).

Atmometer: An instrument for measurement of evaporation rates; also called atmidometer or evaporimeter.

Atmospheric pressure: The pressure exerted by the atmosphere as a consequence of the weight of the air lying directly above the point. At mean sea level atmospheric pressure is normally equal to 76 cm of mercury column.

ATP (Adenosinetriphosphate): Adenosine-derived nucleotide that is the primary source of energy for plants through its conversion to ADP.

Attenuation: The decrease in the intensity of radiation caused by the absorption and scattering of the radiation as it passes through the matter.

Audible voice: A voice, which is neither too loud nor too soft, will attract attention of the trainees and make them comfortable.

Audio materials: Audio materials are those which can be heard only.

Audio-visual materials: The audio-visual materials are those aids that could be heard and seen.

Auditory learners: Learners who learn by listening and respond well to teaching method like lecturing.

Auger (soil): A tool boring holes in the soil. A soil auger is used to withdraw a small soil sample for observation.

Augmentation cropping: Growing of crop to supplement the yield/monetary returns occurring from the main crop. For example growing of forage brassica to augment the fodder yield of berseem in first cut.

Aus rice: Derived from the Bengali word 'Ashu' means *early*. Aus rice is a group of periodically fixed maturing rice varieties that are photoperiod insensitive. This rice is usually direct seeded in March-April with pre-monsoon showers and harvested in July-August, middle of the monsoon season.

Autogamy: (*i*) Pollen grains of a flower pollinate the same flower (*Syn.* Self-pollination).

(*ii*) Self-fertilization; union between male and female gametes produced by the same individual; in *Paramecium*, sexual reproduction within a single animal without conjugation.

Autonomic (vegetative) parthenocarpy: A type of parthenocarpy in which fruits develop without pollination and fertilization.

Autopolyploid: Escaping of a susceptible host plant from damage due to an insect pest, usually, as a result of the plants being at a much less susceptible development stage when the pest population is at its peak.

Autoradiograph: A photograph prepared by labelling a substance with a radioactive material and then allowing the radioactive decay in the substance to produce its image on a photographic film.

Autosome: Chromosomes other than sex chromosomes; their kind and number is the same in males and females of a species.

Auxotroph: A mutant which is incapable of synthesizing an essential biochemical is produced by the wild type individuals of the species.

Available moisture: It is the moisture range or limit available to plants which lies between field capacity and wilting point of a soil.

Available nitrogen: A small proportion of relatively large amount of N present in the rooting zone of soil which is directly available to plants. It is mainly in the form of NO_3 present in soil solution and the dissolved and absorbed NH_4.

Available nutrient: The proportion of any element or compound in the soil that can be readily absorbed and assimilated by growing plants. Available should not be confused with exchangeable.

Available phosphoric acid: The part which is soluble in water or in a week dilute acid such as 2 per cent citric acid.

Available water: The portion of water in a soil that can be readily absorbed by plant roots considered by most workers to be that water which is held in the soil against a pressure of up to approximately 15 bars moisture which lies between field capacity and wilting point.

Axil/Axillary: The angle between the leaf and the stem; hence axillary flower or bud.

Azolla: Azolla is a form of bio-fertiliser which is a floating aquatic fern with nitrogen fixing blue green algae (Anabaena azolla) associated with it, in cavities in the leaf surface. The merits of Azolla-Anabena association lies in its ability to fix nitrogen and also to photosynthesise. It also gives high yields of biomass which can be used as green manure particularly if Azolla is grown in rotation with rice or even in association with rice.

Azospirillium: A non-symbiotic bacterial inoculant that grows around the roots of grasses and other graminaceous crops where it fixes atmospheric nitrogen. Particularly useful in crops which receive little nitrogen fertiliser like nutritious curds (coarse curds).

Azotobacter: A non-symbiotic nitrogen fixing bacterium in the soil utilizing soluble carbohydrates, strongly aerobic susceptible to phosphate deficiency.

B

Back pruning: Cutting back of canes to one to two-bud level for vegetative growth.

Backcross: A cross between a hybrid and one of its parents. Also to make such a cross.

Backcross breeding: A breeding method based on repeated backcrossing of the *F1* and the subsequent generations to the recurrent parent.

Backing: The changing of the wind in a counter-clockwise direction.

Backslope: The slope component that is the steepest, straight then concave, or merely concave middle portion of an erosional slope.

Bacteria: Microscopic one-celled organisms, many of which play an important role in the soil by breaking down organic matter and in animal nutrition by breaking down cellulose into digestible nutrients.

Bacteriophage: A virus that infects a bacterium; usually the host cells are lysed by most bacteriophages.

Badland: An area generally devoid of vegetation and broken by an intricate maze of narrow ravines, sharp crests, and pinnacles resulting from serious erosion of soft geologic materials. Most common in arid or semi-arid regions.

Bagasse: The mill residues from the cane sugar industry consisting of crushed stalks from which the juice has been extracted.

***Bahar* treatment:** Regulation of flowering and fruiting in fruit crops, which flower more than once a year, by cultural, mechanical and chemical means.

Bajada: Coalescing adjacent alluvial fans; a bajada is almost a continuous slope of alluvium along the side of a mountain.

Balanced fertilizer: A soil additive containing suitable proportion of each necessary mineral element to grow a plant or a crop.

Balanced lethals: Lethal genes linked in repulsion phase; they are maintained in this phase either due to tight linkage or crossover suppression; only their heterozygotes survive.

Balanced polymorphism: Two or more alleles of a gene maintained in the same breeding population.

Balausta: A type of fruit which is a modified form of berry, inferior, many celled and many seeded with leathery pericarp; the carpels are placed in two rows, one above the other with thin yellow papery partition walls; the seeds have succulent seed coat (aril) which forms the edible part.

Ballast elements: Elements such as aluminium and silicon have been called ballast elements because they are generally present in large amount, although the plant can grow normally without them.

Band placement: Fertilizer placement on one side or both the sides of the row, about two inches below the seed and two inches of one side of the seed or plant grown, or around each plant.

Band seeding: Placing the seed in rows directly above not in contact with a band of fertilizer.

Band treatment: Treating a narrow strip directly over or in the crop row.

Bar: A term used in a generic sense to include various types of submerged or exposed embankments of sand and gravel built on a sea or lake floor by waves and currents.

Bar chart/diagram: The graphical representation of frequencies or magnitudes by rectangles drawn with height proportional to the frequencies or magnitudes concerned.

Bark: The skin or cover of the stem.

Baroclinic: A baroclinic atmosphere is one in which surfaces of pressure and density intersects at some level or levels.

Baroclinic instability: A type of dynamic instability, associated with a strongly baroclinic region of the atmosphere.

Barograph: A self-recording barometer in which continuous record of the atmospheric pressure is made. The record is called Barograph.

Barometer: An instrument used for measuring the atmospheric pressure.

Barotrophic: A Barotrophic atmosphere is that hypothetical atmosphere in which surfaces of pressure and density coincide at all levels.

Barotrophic wave: A wave-like disturbance in a barotrophic atmosphere.

Barrage: A large structure erected across a river in order to store water, usually for irrigation.

Barren: An area devoid of vegetation.

Barrier: A barrier is an obstruction built across a slope to prevent runoff.

Barrier, mechanical: Mechanical barriers are used as a form of land treatment.

Barrier, vegetative: A method of land treatment wherein vegetative strips or grasses are planted along the contours. These strips of vegetation promote soil building and reduce the velocity of the runoff, thereby providing protection against erosion.

Basal dose: Amount of manures or fertilizer applied to the soil just before the crop is sown or planted or transplanted.

Base: The bottom or lowest part.

Base analogue: Compounds with slight differences from the natural purine or pyrimidine bases; they are incorporated in DNA and RNA.

Base exchange capacity: The extent to which the soil can adsorb exchangeable cations other than the hydrogen and aluminium.

Base saturation percentage: The extent to which the adsorption complex of a soil is saturated with exchangeable basic cations (cation other than hydrogen and aluminium). It is expressed as a percentage of the total cation exchange capacity.

Baseline: An imaginary line (usually marked by a chain) that denotes the longest distance from ridge to valley within a watershed.

Basic chromosome number: It refers to haploid chromosome number (x) of diploid species (2n) or haploid chromosome number of the ancestral diploid species of a polyploid.

Basic number: The haploid chromosome number of the ancestral diploid species of a polyploid. Represented by x, also haploid chromosome number of diploid species. X chromosomes constitute a genome.

Basic number: The haploid or gametic chromosome number of a diploid species (*genomic number*); in polyploid species, haploid number of the parental diploid species; represented by *x*.

Basic seed: Seed produced through mass selection (with progeny test) in a pure line variety or clone; source of breeder seed.

Basic slag: A by-product of steel industry obtained from phosphate iron ores. It contains about 6 to 18 per cent phosphoric acid.

Basin: A basin is an area drained by a single river system. The term is also used for the hollow ridge that is made around a tree.

Basin irrigation method: In this method, water is applied in small plots or nearly flat formed in the field by ridges.

Basipetal: Towards the base of a plant organ; generally downwards in shoots and upwards in roots. Opposite of acropetal.

Basophiles: The plants which can grow in alkali soils (*p*H range 7.4-8.5); also known as alkali plants.

Baul unit: Unit of fertilizer or any other growth factor, be taken as that amount necessary to produce a yield of 50 per cent of the maximum possible.

Beaches: Sandy, gravelly, cobbly, or stone shores washed and reworked by waves. The areas may be partly covered by water during high tides or storms.

Bed planting: Planting on elevated level beds which are separated by furrows.

Bedder: A tillage tool to form ridges or narrow beds.

Bedding: The mounding of soil into elevated strips performed with tillage tools such as sweep, shovel, disk, or moldboard.

Bedrock: The solid rock underlying soils and the regolith in depths ranging from zero (where exposed by erosion) to several hundred centimetres.

Bench terrace: Constructed to make sloppy land cultivable or stable and consists of a series of platforms or mostly level benches cut into the hill slope in a step like formation; the platforms being separated by very steep sides by rock or vegetation.

Bench terracing: It consists principally of transforming relatively steep hill slopes into level strips thereby reducing the degree of slope considerably. It is practiced mostly in rainfed areas with high rainfall, good soil depth and a slope range between 16 per cent-30 per cent, since such areas are prone to the scouring action of the runoff water. A bench consists of a batter (also known as riser/vertical drop) made up of rock or earth, protected by a heavy growth of grasses. The batter separates one bench from the next. There are three types of bench terraces: (*a*) **Level or table top** which are suitable for areas with deep permeable soils and an even distribution of rainfall, (*b*) **Sloping outward** are suitable for areas with permeable soils of a medium depth and less than 600 mm rainfall. The major function of this terrace is to distribute the scare rainwater across the entire bench. **Sloping inward** is suitable for areas with soils of poor permeability. Good to medium soil depth and a rainfall beyond 600-750 mm. These terraces help in draining the excess water through the drains.

Benchmark: A point of known elevation, used as a reference point against which the different levels of an area under a land survey are measured. There are four types of benchmarks: (*a*) **GTS (Great Trignometrical Survey)** Benchmarks are established with high precision at intervals all over the country, by the Survey of India. Their positions are indicated on maps. The mean sea level at Karachi is taken as the standard data against which all the other levels are marked Benchmarks. (*b*) **Permanent Benchmarks** are points of reference fixed by the Public Works Department (PWD) on clearly defined and permanent structures like parapet walls of bridges or culverts, corners of plinths of building, gates pillars and milestones. (*c*) **Arbitrary Benchmarks** are taken on permanent objects whose elevation levels is assumed arbitrarily. This is done when GTS or permanent benchmarks are not available in the vicinity. Arbitrary benchmarks are used in small areas (100-500 ha) in secluded pockets where it is not critical to connect contour values to the contours of adjoining areas, (*d*) **Temporary Benchmarks** are reference points established at the end of a day's work when there is a break. They are also established on relatively permanent structures like gates, stones, rocks etc. Work when resumed is continued from these points.

Beneficiary Index Card: All the landless labourers in a watershed are provided with a beneficiary index card. It is one of the devices proposed to facilitate monitoring by the labourers themselves. It serves as a record of the activities/benefits realized by the landless labourers in a given micro watershed.

Bentonite: A highly plastic clay consisting of the minerals montmorillonite and beidellitre (smectite), which sweel extensively when wet; also called volcanic clay, soap clay, and amargosite.

Berm: It is the term used to refer to a 1.52 m feet wide strip, built across a slope on the downstream face of a earthen dam. The width of the strip varies with the type of soil. *For example*: A berm constructed of red soil requires lesser width than one constructed of black soils. There are two types of mechanical barriers: Contour bunds and Graded bunds.

Berry: A simple fruit derived from one flower, in which a parts remain succulent. The true berry is derived from an ovary (grape); the false berry from an ovary plus receptacle tissue (**blueberry**).

Beta oxidation: If out of 4 carbon atoms of butyric acid 2 carbon atoms are cleaved, only 2 atoms are left in the remaining acetic acid. This process is known as beta oxidation as in case of conversion of 2,4Dβ into 2,4-D.

B-horizon: The subsoil layer in which certain leached substances are deposited.

Biennial: A plant that completes its growth in two years. In the first year it produces leaves and stores food; in the second year it produces fruits and seeds.

Biennial bearing: It refers to the pattern of bearing in which normal commercial crop load "on year" is followed by "off year", with no or negligible crop.

Binding energy: The energy with which a particle is held to an atom or nucleus.

Binomial distribution: A discrete probability distribution expressing the probability of one set of dichotomous alternatives that is success or failure. Also known as the Bernoulli distribution.

Binomial Expansion: Exponential multiplication of an expression consisting of two terms joined by a + or a – sign, *e.g.*, $(a+b)^n$ or $(a-b)^n$

Bioassay (biological assay): (*i*) Determination of the relative strength of a substance (hormone, drug, etc.) by comparing its effect on a test organism with that of a standard preparation.

(*ii*) A method for the quantitative determination of a substance by its effect on the growth of a suitable microorganism, plant, or animal under controlled conditions.

Biochemical antagonism: The type of antagonism that occurs when one agrochemical (the antagonist) decreases the amount of a given herbicide that would otherwise be available to its site of action in the absence of the antagonist.

Biochemical genetics: The genetics of differences in the ability or inability to synthesize or metabolize certain chemicals.

Biochemical Oxygen Demand (BOD): Biochemical oxygen demand (BOD) the most commonly used parameter to define the strength of a municipal or organic industrial waste loadings to treatment plants and in evaluating the efficiency of such treatment systems. BOD is by definition the *quantity of oxygen utilized by a mixed population of microorganisms in the aerobic oxidation* (of the organic matter in a sample of waste water) at a temperature of 20°C.

Bioconcentration: It is the ability of microorganisms to store materials in the environment. The storage may be within the cells, as in the case of sulphur, or outside the cells, as in the case of nitrate or sulphur.

Biocontrol: Use of biological agents, *e.g.,* viruses, bacteria, fungi, etc. for the control of pathogens, insect pests, weeds etc.

Bio-converson: A general term describing the conversion of one form of energy into another by using like animals, plants or micro-organisms. Synthesis of organic compound from carbon dioxide by plants is bio-conversion of solar energy into stored chemical energy.

Biodetradable: A substance capable of being decomposed by biological processes.

Biodiversity: The variety of living things expressed in genetic, species and ecosystem dimensions. There may be more species terms generated, *e.g.*, agro-biodiversity for the biodiversity of agriculture; cultural diversity for the biodiversity of human culture.

Biodynamic farming: A holistic system of agriculture devised by Rudolph Steiner that seeks to connect nature with cosmic creative forces. An attempt is made to create a whole-farm organism in harmony with its habitat. Compost and special preparations (*e.g.*, plant-derived sprays) are used. Synthetic fertilisers and pesticides are avoided.

Biofertiliser: Fertiliser containing micro-organisms, the action of which renders the minerals available to plants, *e.g.*, rhizobia, azotobactor, PSB etc.

Bio-gas: It is the term used to describe gaseous mixture of methane and carbon dioxide produced by the fermentation of organic wastes. The wastes include: animal dung, night soil, poultry droppings, sheep and goat droppings, kitchen waste, daily wastes, vegetative stubble, wood waste, sugarcane waste, press muds and sawdust. Gas production normally takes about 15 to 20 days. Bio-gas is a good substitute for cooking gas and electricity. The by-product is used as a natural fertilizer.

Biological control (pests): The use of natural enemies to control pests, including both controlled with imported natural enemies and augmentation and conservation introduction, augmentation, and mass release of natural enemies parasites, predators and disease organisms through manipulation of the pest host, the environment and/or the enemies themselves (IASA 1990).

Biological half life (radio tracer): The time required for one-half of an administered substance to be excreted from the body or from an organ of living tissue.

Biological magnification: The accumulation of increasing amounts of a substance in organisms of a food chain as one progresses up the chain (*i.e.*, low in plankton, higher in small fish, and highest in fish-eating birds).

Biological mineralization: The conversion of an element occurring in organic compounds to the inorganic form as a result of biological decomposition.

Biological pesticides: Biological entities or their products used to control insects pests, *e.g., Bacillus thuringiensis* spores or a preparation of the crystal protein produced by it.

Biological value of proteins: The degree of usefulness of a protein in human nutrition.

Biological weathering: Refers to physical and chemical weathering assisted by biological agencies.

Biological yield: Refers to all dry matters produced by a crop in a unit land area.

Biomass: (*i*) The total mass of living microorganisms in a given volume or mass of soil. (*ii*) The total weight of all microorganisms in a particular environment. (*iii*) Any organic matter which is available on a renewable basis, including agricultural crop and agricultural wastes and residue, wood and wood wastes and residue, animal wastes and aquatic plants, usually expressed as dry weight per unit area.

Biome: Easily recognised community formed by the interaction of regional climates with regional biota and substrates. In a given biome the life form of the climatic vegetation is uniform. Thus the climax vegetation of the grassland biome is grass, although the dominant species of grass may vary in different parts of the biome.

Biometry: The application of concepts and methods of statistics to the study of biological problems.

Biosensor: Biosensors are the combinations of biologically active material displaying characteristic specifying with chemical or electronic sensors to convert the response into electrical signal. They are, in fact, biocatalysts and are used as purified enzyme, antibody or as a whole microbial cell are as an organelle. The response of biosensor is measured in terms of substrate used or product form. They are of different types like, carbon electrode, thermister, aednosine electrode etc.

Biosequence: A sequence of related soils that differ, one from the other, primarily because of differences in kinds and numbers of plants and soil organisms as a soil-forming factor.

Biosequuam: One sola above another in the same profile.

Biosphere: The earth's envelope in which living organisms exist in their natural state. It includes atmosphere, animals, vegetation, soil and oceans.

Biotechnology: The application of scientific engineering principles of processing of materials by biological agents to provide goods and services.

Biotic factors: Living components of environment, which include microorganisms, plants of the same or of other species and animals including man.

Biotransformation: Conversion of a less useful compound into a more valuable and useful one through the action of an organism or (more precisely) by an enzyme produced.

Biotype: Strains of a species of pathogen, particularly an insect pest, differing in their ability to attack different varieties of the same host species (*Syn.*, physiological races).

Biparental cross analysis: A design for genetic analysis based on progenies derived from the matings involving a random samples of *F2* or later generation plants from a cross between two pureline; four designs based on such progenies are NCD I, NCD II, NCD III, TTC.

Bisexual: Bearing both male and female reproductive organs by the individual, especially in the plant.

Bitter pit: A disorder of apple caused by calcium deficiency which is characterized by small, brown, necrotic zones in the flesh 3-5 mm in cross-section, more frequent towards the calyx portion of the fruit and sometimes visible through the skin as dark green or brown in flesh.

Biuret: A condensation product of two molecules of urea and is toxic to many plants. Its content in the commercial fertilizer should not exceed 0.2 per cent.

Bivalent: The structure formed by pairing of two homologous chromosomes during meiosis; seen during pachytene onward till the beginning of anaphase I.

Black end/hard end of pear: A physiological disorder of pear fruit, associated with drought caused by certain rootstocks and is characterized by blackening and hardening of tissues surrounding the calyx opening. Rootstocks *Pyrus pyrifolia* and *P. ussuriensis* causes it.

Black heart: Darkening of inner areas of woody stems by deposits that prevent conduction or food storage caused by death of cells from winter freeze damage.

Black soils: They vary from shallow to deep. The typical soil derived from the Deccan trap is the regur or black cotton soil common in Maharashtra, western M.P. and Gujarat. Black soils are highly fertile except in the uplands. They are fine grained, dark and contain a high proportion of calcium and magnesium on the clays. They have high moisture retention capacity and are sticky when wet. Owing to considerable contraction on drying large cracks are formed. These soils contain abundant iron and fairly large quantities of lime, magnesia and alumina. Potash is also found. They are poor in phosphorus, nitrogen and organic matter.

Black tip: A disorder of mango caused due to brick kiln fumes and is characterized by blackening and hardening of the distal end of fruits which ripe prematurely.

Blind cultivation: Cultivation with a harrow weeder, rotary weeder or other implements to kill weeds before a seeded or planted crop has come up.

Blind hoeing: It is the hoeing which is done when the crop has not sprouted above the ground level on the sown buds or eyes are not up. This type of hoeing is done in sugarcane for removing weeds before the eyes have come up.

Blossom end rot of grape: A disorder due to calcium deficiency characterized by a black sunken spot develops at the blossom end of the berry which later on spreads with water soaked region around it.

Blotch: A disease showing large and irregular spots on stem, shoots and leaves.

Blown-out: Areas from which all of the soil material has been removed by wind erosion. Usually barren, shallow depression with a flat or irregular floor consisting of a more resistant layer and/or accumulation of pebbles, or a wet zone immediately above a water table. Usually unfit for crop production.

Blue green algae: An heterogeneous group of prokaryotic photosynthetic organisms which contain chlorophyll. They include unicellular colonial and filamentous species. They are obligate phototrophs and store cyanophcean starch. They are non-symbiotic, free-living Nitrogen fixers which grow very well in water logged soils like in rice fields. Beside fixing atmospheric Nitrogen, Blue-green algae secrete vitamins and growth promoting substances which contribute to better growth of the rice plants.

Bole: The main stem of a tree and sometimes only the lower part of the stem up to a point where the main branches are given off.

Bone ash: It is obtained by burning bones with free access to air and contains 30-40 per cent P_2O_5.

Bonemeal (raw): A fertilizer made of dried animal bones finely ground. It contains 20 to 24 per cent P_2O_5 of which 16 per cent is citrates soluble. The availability of its plant food depends largely upon how fine it is ground. It contains at least 3 per cent nitrogen too.

Boot stage: Growth stage of grasses at the time the head is enclosed by the sheath of the uppermost (flag) leaf.

Borax: A chemical compound that contains approximately 11 per cent boron. Borax is applied to boron-deficient soil or sprayed on the plant's foliage which indicate boron deficiency.

Border Irrigation (border strip irrigation): It is an efficient method of irrigation for close growing crops and consists essentially of dividing the field by low flat levels into series of strips, each of which is flooded separately. Generally, these strips extended in the direction of the steepest slope (not more than 3 per cent) and at right angles to the supply ditch. The strips are made level transversely but follow a gentle down field slope longitudinally.

Boro-rice: Winter season rice transplanted in December-January and harvested in April-May. Boro is photoperiod insensitive and grown under irrigated conditions.

Borrow pit: During the construction of an embankment, earth is excavated from the upstream side of the slope. The depression that is formed as a result of this activity is termed as a borrow pit. The pit fills up in course of time with the accumulated sediment along with the runoff.

Botanical pesticide: A plant-derived pesticide.

Bottom heat: A technique for rooting of cuttings which entails the application of artificial heat to the rooting medium to keep the basal end of the cutting comparatively warm for inducing rooting.

Boundary based bunding: It is an indigenous soil and water conservation technology evolved by farmers. It is especially useful for farmers with small land holdings. In this system, the bunds are constructed along the boundary and therefore do not take away extra land from cultivation as contour bunds do. There are two types of boundary-based bunds. (*i*) Field bunds used in areas with-rainfall below 500 mm, permeable soils and 500-600 mm rainfall. They help in conserving rainwater and soil. (*ii*) Field drains used in areas with rainfall above 600 mm, impermeable soils and rainfall of 500-600 mm rainfall. They help in conserving the soil and draining away excess moisture.

Boundary layer: That layer of fluid adjacent to a physical boundary in which the fluid motion is much effected by the boundary and has a mean velocity less than the free-stream value.

Bower system: A system of training grapevine in which the vine is pinched off after reaching the wire network and two vigorous shoots in opposite direction are selected as primary arms. On each primary arm, three secondary arms are allowed to grow and have about eight to ten tertiaries which after maturity form fruiting canes.

Bracing: A method of strengthening branches that form weak crotches in which a small branch to be supported by bark or bridge grafting and sometimes small branches from the two limbs, are twisted together and allowed to unite naturally.

Brackish: Slightly salt. Term applied to water with a saline content between 15 and 30 parts per thousand; seawater has more than 30 parts per thousand.

Bract: A special leaf or leaf like structure, usually at the base of a flower or inflorescence.

Brain storming: Brain storming is a group creativity technique designed to encourage lateral thinking around a stated question or problem.

Bread wheat: Most common grown wheat, a hexaploid of the group *Triticum aestivum.*

Breba: The first or spring crop of fig.

Breccia: A rock composed of coarse angular fragments cemented together.

Breeder seed: Seed produced by the breeder or the institute which developed the variety; source of foundation seed.

Bridge grafting: A grafting method in which one year old scion above and below the injured areas of trunk or limbs are inserted to save the tree. The scions are usually collected during the dormant season and are inserted in the tree when the bark begins to slip.

Bridging species: A species used in gene transfer from one species to another sexually incompatible species; bridging species is compatible with both the donor and recipient species; it may be a natural or synthetic species.

Brine solution: Salt water or solution.

Brix: The percentage of total solids sugarcane juice, read from brix hydrometer.

Broad leaf (weeds): Used in weed terminology to designate a broad group of non-grass like plants, usually dicots.

Broad-base terrace: A ridge type terrace 25-50 cm high and 5-10 metres wide with gently sloping sides, a rounded crown and dish shaped channel along the upper side, constructed to control erosion by diverting run off along the contour at non-scouring velocity.

Broadcast treatment or blanket application: It is the term employed for the application of herbicide uniformly to an entire area.

Broadcasting: The process of scattering of agricultural inputs, such as seed, fertilizer and manure on the surface of the soil by, and/or by implements to provide uniform distribution of materials over the entire field.

Bromelin: A protein digestive enzyme present in the mature pineapple fruit.

Brown core: A low temperature disorder in apple where browning and necrosis of flesh occur around the seed cavity and in the flesh underlying the stem cavity.

Brown staining: A type of chilling injury in grapefruit and mandarins where diffuse irregular superficial discoloration of the peel occurs.

Browning of litchi: Discoloration of the pericarp associated with desiccation of fruit and development of off flavour after harvest.

Brush Control: Control of woody plants.

Bud: A plant structure that contains an underdeveloped short or flower.

Bud mutation: Mutation in somatic tissues, usually affecting an axillary bud.

Bud nicking: A deep knife incision through the vascular strands into the wood directly below a bud to retard its growth.

Bud notching: Removal of a wedge shaped piece of bark with wood directly above a bud to stimulate growth.

Bud selection: A form of clonal selection in which mutant buds are selected.

Bud sport: A new strain or clone arising from a mutated bud.

Budding: A form of sexual reproduction in which a new individual develops as a growth of a mature organisms.

Buffer compounds, soil: The solid and solution phase components of soils that resists appreciable *p*H change in the soil solution (*i.e.*, carbonates, phosphates, oxides, phyllosilicates, and some organic materials).

Buffer strips: Contour strips of grass or other erosion resisting vegetation between or below cultivated strips of fields.

Buffering: (*i*) The capacity of the solids and liquids to resist appreciable change in *p*H of the soil solution. (*ii*) The ability to maintain the approximate concentration desired to any ion in the soil solution.

Buffers: A buffer solution is one which is resistant to changes in *p*H on the addition of small amount of acid or alkali. This resistance is called the buffer action. The magnitude of the buffer action is called the buffer capacity.

Bulb: An underground stem with fleshy, food-storing scale leaves, essentially a below ground bud.

Bulk density: It is the ratio of the mass of a soil's solids (oven drip soil) to the bulk volume of the soil. It is expressed as mass per unit volume, generally g/cc.

Bulk density (loose): The mass per unit volume of a material after it has been tipped freely into a container under clearly specified conditions.

Bulk density (tapped): The mass per unit volume of a material tipped into a container and then compacted under clearly specified.

Bulk fertiliser: Commercial fertilizer delivered to the purchaser in the solid state, in a non-packed form.

Bulk method: In this method, *F2* and the subsequent generations are grown in bulk, usually without artificial selection. In the end, pureline varieties are developed through individual plant selection.

Bulk volume: The volume of an arbitrary soil mass including the volume of the solid particles and of the pores (interstices, voids).

Bulky organic matter: These manures are bulky in nature and supply plant nutrients in small quantities and organic matter in large quantities.

Bun: Method of cultivation in Meghalaya which involves putting of dried vegetation in the form of raised beds along the slopes and covering the same with soil collected from the surroundings, burning of the covered vegetation and planting on the soil afterwards.

Bund: An artificial earthen embankment made across sloping agricultural land to cut short lengthy soil slopes and reduce run-off and erosion.

Bund former: An implement used for making bunds or ridges by collecting the soil. Bunds are required to hold water in the soil thereby conserve moisture and prevent run off.

Buried soil: Soil covered by an alluvial, loessal, or other depositional surface visibly when treated with cold 0.1 M HCl. These soils usually contain from as little as 10 to as much as 200g Kg^{-1} $CaCO_3$ equivalent.

Burr knots: Rough or irregular protuberances on the stem caused due to preformed latent roots (root initials) as seen in some cultivars of apple and quince.

Bush-fallow system: Farming systems in which the natural regeneration of self-propagated plants in successional community is the restorative agent in respect of nutrients, organic matter, water conservation and microclimate. Crop plants are grown on patches cleared by felling of trees and or burning of grasses and herbs; fallow succession dominated by woody shrubs and grasses.

Button shedding: Shedding of immature nuts in coconut palms due to lack of pollination, nutritional imbalance hormonal imbalance and pest and disease attack.

Buttress root: A plant-like outgrowth at a tree's base, providing extra support to the tree.

C

'C' horizon: A mineral horizon generally beneath the solum that is relatively unaffected by biological activity and pedogenesis and is lacking in properties diagnostic of an 'A' or 'B' horizon.

C_3 plant: A plant in which the first product of CO_2 fixation is the 3-carbon compounds; phospholyceric acid and is comparatively.

C_4 plant: A plant in which the first product of CO_2 fixation is the 4-carbon compound oxalo acetic acid and is photo-synthetically more efficient than C_3 plants, *e.g.,* Sugarcane, Maize etc.

Cadastral map: A large scale revenue map of a village. It contains information on all aspects related to the topography of the village—the presence of roads, wells, wetlands, drylands, habitation, outcrops, tanks, streams, rivers etc.

Caking: The formation of coherent mass for individual particles.

Calcareous: Composed of or containing calcium carbonates ($CaCO_3$).

Calcareous soil: Soil containing sufficient calcium carbonates (often with magnesium carbonate) to effervescence visibly when treated with cold 0.1 N hydrochloric acid.

Calchic horizon: A diagnostic mineral horizon of carbonate accumulation indicated by letter K.

Calcification: Process including accumulation of calcium carbonate.

Calcifuge: Plants suited to grow on acid soils.

Calcitic lime: Limestone containing mostly $CaCO_3$.

Calcium: It is observed as divalent Ca^{2+} easily available for plant growth, but acidic soils where high rainfall occurs are often fertilized with lime (a mixture of CaO and $CaCO_3$) to raise the *p*H.

Calcium ammonium nitrate (CAN): A commercial nitrogenous fertilizer, consisting of ammonium nitrate and powdered limestone or dolomite, containing 20 per cent nitrogen. One half of the nitrogen is in the literate form and the remaining half in the ammoniacal form.

Calcium cynamid [$CaCN_2$]: A commercial product consisting principally of calcium and carbon. It contains not less than 20 per cent of nitrogen, which is not immediately plant-available.

Calcium metaphosphate [$Ca(PO_3)_2$]: A product obtained by treating phosphate rock with gaseous phosphorus pentoxide (P_2O_5) at high temperature.

Calcium nitrate [$Ca(NO_3)_2$]: The calcium salt of nitric acid. It is an excellent source of nitrate form of nitrogen and of water-soluble calcium. The commercial product contains about 15 per cent nitrogen and 28 per cent lime (CaO).

Calibration: Series of operations to determine the amount of solution applied per unit area of land and the amount of chemical required for a known volume of diluent.

Caliche: (*i*) A zone near the surface of, more or less cemented by secondary carbonates of Ca or Mg precipitated from the soil solution.It may occur as a soft thin soil horizons, as a hard thick bed, or as a surface layer exposed by erosion. (*ii*) Alluvium cemented with $NaNO_3$, NaCl and other soluble salt in the nitrate deposits of Chile and Peru.

Callus: The undifferentiated mass of parenchyma tissue at the edges of a bud, graft or *in vitro* cultured explants.

Calm: Absence of appreciable wind; On the Beaufort Scale of the wind force calm accorded the figure 0 and has a wind speed equivalent or less than 1 knot.

Calmodulin: Calmodulin has been isolated from lower and

higher plants, the green algae Chlamydomonas and Mougetia; the fungi, *Neurospora crassa.* The yeast *candida elbucans* and *Saccharomyces cervisaiae. Blastocladia emersonji*, two *Basidiomycetes* and different sources of higher plants including peas, mungbeans, wild carrots and peanut seeds, barley shoots and roots, *Zucchi hypocotyls*, spinach leaves, corn etc.

Calorimeter: An apparatus used for measuring the amount of heat generated or emitted by any substance.

Cambium: A meristem with products of divisions arranged in an orderly fashion in parallel files. It consists of one layer of initial cells and their derivations.

Cane sugar: A type of storage sugar (sucrose) present in sugarcane, sugar beet, ripe fruits etc.

Capillary water: In that which is held by the surface tension forces as a continuous film around the soil particles or water retained within soil pores.

Caprification: The process of pollination of the edible fig (smyma and san pedro types) with pollen from the caprifig by the *Blastophaga psens*, wasp.

Caprifig: The wild or uncultivated fig used to pollinate the domestic edible fig.

Capsule: A dry fruit formed out of two or more fused capsules, which split open when ripe.

Capture fisheries: Exploitation of fishery wealth of natural water bodies without attempting to rear fish. Multiplication of species occurs naturally.

Carbohydrates: Compounds of carbon, hydrogen and oxygen in the ratio of one atom each of carbon and oxygen to two of hydrogen, as in sugar starch and cellulose.

Carbon: The special role and function of carbon in plant life as being essential elements. The uptake is mainly through the roots in the liquid phase, but carbon uptake is mainly through the leaves and shoots in the gas phase. The toxicity symptoms of CO_2 in mature tomato and bean leaves, yellowing spots, bulkness etc. resemble the signs of deficiency of P and toxicity of Mg and could be related to the accumulation of starch.

Carbonation: A chemical weathering process combining the action of acidity and formation of soluble bicarbonates. The solvent is carbonic acid formed from carbon dioxide dissolved in water.

Carbon-nitrogen ratio: The ratio of the weight of the total organic carbon to the weight of total nitrogen in a soil or in an organic material, C:N Ratio of wheat straw is nearly 80:1 while that of soil is 10:1 when undecomposed straw with a high C:N ratio is applied to the soil its C:N ratio is reduced through bacterial decomposition. To speed up bacterial decomposition, nitrogenous fertiliser is added to the soil at the time of turning under straw.

Career training: Training designed to update the knowledge, skills and abilities of employees to help them assume greater responsibility in higher positions.

Carotene: An orange pigments occurring in certain plastids and is the precursor of vitamin A. An individual heterozygous for a recessive, usually detrimental, gene.

Carrier: The liquid or solid material added to a chemical compound to facilitate its storage, shipment and application.

Carrying capacity: The maximum numbers of users that can be sustained by a given land resources at a particular level of technology.

Caryokinesis: The division of a nucleus into two daughter nuclei; through mitosis or meiosis.

Case study: A case study consists of materials related to an event, or problem. Since most of the case studies are related to real life situation, the analysis of case studies will help in acquiring development of analytical skills.

Cash crop: A high value marketable crop, *e.g.,* sugarcane, cotton, jute, tobacco, tea etc.

CAT: Abbreviation of Clear Air Turbulence.

Cat clays: Strongly acidic clays resulting from aeration of anaerobic soils containing many sulphide minerals. The sulphides are oxidized to sulphuric acid in aerated soil.

Catalyst: Substance that accelerates the rate of chemical reaction without being used in the process.

Catch crop: A quick growing crop incidentally planted and harvested between two major crops in consecutive seasons, also it may be a contingency crop grown to replace a major crop which has failed.

Catchment area: Catchment Area is the term used to describe the entire area under a watershed that drains into a source. An area from where entire rainwater is collected and then conducted to a reservoir.

Cation: An ion carrying positive charge of electricity. The common soil cations are calcium, magnesium, sodium, potassium and hydrogen.

Cation-exchange capacity: The sum total exchangeable cation absorbed by a soil, expressed in milli-equivalent per 100 g. of soil or CMOL (P^{+}) kg^{-1}. Measured values of cation-exchange capacity depend somewhat on the method used for the determination.

Catkin: A crowded spike of tiny flowers, hanging down from the twigs of the plants, which bears unisexual flowers only.

Cauliflory: Bearing of fruits directly on the branches or main stem.

Ceiling leaf area index: Leaf area index at which mutual shading of leaves in a canopy is complete and net photosynthesis is zero.

Cemented (indurated): Having a hard, brittle consistency, even when wet, because the particles are held together by cementing substances, such as humus, calcium carbonate, or oxides of silicon, iron and/or aluminium.

Centres of diversity: Areas where cultivated plant species and/or their wild relatives show much grater variation than anywhere in the rest of the world.

Centres of origin: Areas where cultivated plant species are supposed to have originates. Based on centres of diversity.

Centromere: The localised region of a chromosome where spindle fibers attach; involved in chromosome movement during cell division; seen as a constriction in each mitotic metaphase chromosome.

Cereal forage: A cereal crop harvested when immature for hay, silage, and green feed or as pasturage.

Cereals (crops): Cereals are the crop plants belonging to the grass family graminaceae which are grown for their edible starch seed botanically known as caryosis. The word 'Ceres' means the goddess of grain, *e.g.,* wheat, rice, maize.

Certified seed: Seed produced form foundation, registered or certified seed. Its purity is certified by a seed certification agency, and is usually used for commercial crop production.

Chaff: Glumes, husks or other seed covering combined with other plant parts separated from seed in threshing or processing.

Chaffing: Pneumatic separation of very light materials from the produce.

Chaining: A method by which a heavy chain is dragged, along a water channel to remove aquatic weeds.

Character: Derived from characteristics; a morphological, anatomical, physiological or behavioural feature of an organism; usually, the end-product of development; a result of the genotype, environment and genotype interaction effects.

Charle's law: At constant pressure, the volume of a given mass of gas is directly proportional to the absolute temperature.

Check: A Check is a barrier laid across a rill or a gully to control the erosive velocity of runoff water. Checks protect the gully beds by reducing the speed of runoff water, redistributing the runoff water, increasing percolation, encouraging silting and improving soil water regime for establishing a plant cover. A check is not made to the specification of measurement. This makes it different from the other structures used in the treatment of drainage lines. Types of Checks: Live vegetative edges, brushwood dams, boulders, sand bags, brick masonry, mall, and earthen barriers.

Check irrigation (Check basin irrigation): Irrigation method whereby earth ridges are built around that area partly or wholly to retain water to form a pond.

Check row planting: The process of planting in which row to row and plant to plant distance are uniform and plants across the rows are also in line.

Chelates: An inorganic compound capable of holding the plant nutrient in a form which prevents it from getting tied with other elements in the soil, keeping it more or less in more available form for the plant. The term refers to the claws of a crab illustrative of the way in which the atom is held.

Chelation (Key-lay'-shun): The formation of strong bonds between metals and organic compounds. Some are insoluble and help to keep all the nutrient metals (iron, zinc, manganese, and copper) mobile in the soil, thus aiding plant uptake.

Chemical antagonism: The type of antagonism that occurs when an antagonist reacts chemically with a herbicide to form an inactive complex.

Chemical dormancy: Type of seed coat dormancy in which germination inhibiting chemicals (phenols, coumarin and abscisic acid) accumulated in the fruit as well as in the seed covering and strongly inhibit germination.

Chemical Oxygen Demand (COD): A measure of the oxygen-consuming capacity of inorganic matter present in water of wastewater. It is expressed as the amount of oxygen consumed from a chemical oxidant in a specific test.

Chemical weathering: The breakdown of rocks and minerals due to chemical activity, primarily due to the presence of water and components of the atmosphere.

Chemigation: The process where fertilizers and pesticides are applied into irrigation water to fertilizer crops and control pests.

Chemodenitrification: Non-biological process leading to the production of gaseous forms of nitrogen (molecular nitrogen or an oxide of nitrogen).

Chemotaxonomy: A method of classification of plants where the chemical constituents of plants are used in taxonomy.

Chiasma: The point of contact (initially, representing the point of exchange) between nomister chromatids of homologous chromosomes during and after diplotene (till the beginning of anaphase I); produced as a result of crossing over.

Chicle gum: The coagulated resinous latex derived from the bark of sapota used mainly for making chewing gum and in dental surgery.

Chilean nitrate of soda (Chile Saltretre): A product obtained by refining the crude nitrate deposits found in Chile and containing about 99 per cent sodium nitrate.

Chilling injury: An injury caused by relatively low temperatures above the freezing point.

Chilse: A tillage implemented with one or more shanks to which are attached chisel, spike, or narrow-shovel tools. When used to subsoil, and penetrate deeper than 0.4m, the shanks are sturdier and spaced farther apart.

Chimera: An organism or organ having tissues of different genetic constitutions or origin; a consequence of gene mutation, natural or artificial fusion of different zygotes or organs (*e.g.*, grafting).

Chimeras: A plant consisting of cells two or more genotypes in the same part of a plant. It may result from mutation, irregular mitosis and somatic crossing over.

Chip budding: Type of budding in which chip of bark with wood is removed from the stock and replaced by a same size of chip containing a mature bud from desired scion.

Chiselling: Breaking of loosening the soil, without inversion, with a chilse cultivator or chilse plow, usually below normal plow depth.

Chi-square (χ^2) Test of independence: Test to determine whether two attributes are includes pendent by comparison of observed frequencies relative to expected frequencies.

Chi-Square (χ^2) test: A statistical test for determining whether observed date agree with an expected ratio; c^2 $\Sigma[(O - E)2/E]$.

Chlorite: A layer structure group of silicate minerals of the 2 : 1 type that has the interlayer filled with a positively charged metal-hydroxide octahedral sheet. There are both triochtahedral (*e.g.*, M = Fe^{2+}, Mg^{2+}) and dioctahedral (M^+ Al^{3+}) varieties.

Chlorophyll: A light trapping green essential for photosynthesis.

Chloroplast: A cellular structure (plastid made up of protoplasm) and containing the green pigment chlorophyll wherein the photosynthesis is carried out; has DNA and participates in cytoplasmic inheritance.

Chlorosis: A loss of normal green colour of the plant. Colours may be uniform pale green, brown spotted, yellowish over the entire leaf, or yellowish only between the leaf veins.

Chopping: The process of cutting plant materials with knives or other sharp instruments.

Chroma: The relative purity, strength or saturation of a colour; directly related to the dominance of the determining wavelength of the light and inversely related to grayness; one of the three varieties of colour. See also Munsell colour system, hue and value colour.

Chromatid: Substructures of chromosomes produced by replication of preexisting Chomatids. Each chromosomes has two chromatids, which appear fused at the centromere.

Chromatin: A complex between DNA, histones and other proteins, and chromosomal RNA, making up the chromosomes; in form of 300 A *chromatin fibres* (derived from, chromaticity = property of taking up stain).

Chromocentre: In salivary gland chromosomes of certain Diptera; the body produced by a fusion of the heterochromatic regions of the autosomes and *X* chromosome, and the entire *Y* chromosome.

Chromomere: Small bead-like structures present in chromosomes, especially observable during prophase I of meiosis, having characteristic size and location; most likely produced by a more dense packing of chromatin.

Chromosome: (*i*) Nucleoprotein bodies composed of 300 A chromatin fibers; clearly observable during cell division; dark-staining with basic dyes; bearers of the genetic material and the genes.

(*ii*) Thread-like structure present in nucleus, which shows distinct change in their morphology during cell division. They are deeply stained with some dyes, particularly with the Feulgen's reagent. Genes are located in chromosomes in a linear order.

Chromosome Aberration: A change in chromosome number from the diploid state, or a change in chromosome structure from the normal chromosome complement.

Chromosome Banding: The staining of chromosomes following specific procedures which produce many dark and light zones along their lengths; the pattern of banding is highly specific, and all the 23 human chromosomes can be identified.

Chromosome manipulation techniques: Techniques used for promoting gene transfers from chromosomes of a related species into those of a cultivated species.

Chromosome substitution: The procedure of the process of replacing one pair of chromosomes of a variety with those of another variety or a related species. The latter is alien substitution.

Chronosequence: A sequence of related soils that, differ one from the other, in certain properties primarily as a result of time as a soil-forming factor.

Cis-Arrangement: *See, Coupling Phase.*

Cistron: A segment of DNA which codes for one polypeptide; the unit of function; identified by *cis-trans* or complementation test; usually, mutants within a cistron, do not show complementation; generally, *accepted as a synonym for gene.*

Citrate-insoluble phosphorous: That portion of P in fertiliser remaining after water (*see* water-soluble phosphate) and ammonium citrate (*see* citrate soluble phosphonic acid) extractions. Phosphorous content of fertiliser that is considered to be immediately unavailable in plants.

Citrate-soluble phosphoric acid: It is that part of the total phosphoric acid (as P_2O_5) in a fertiliser that is insoluble in water but soluble in neutral N ammonium citrate solution.

Citric acid cycle (Krebs' cycle): A cycle series of reactions, which transforms syruvic acid into dioxide and hydrogen, which then goes terminal oxidation to yield available energy.

Clay (soils): (*i*) A mineral soil separate consisting of particles less than 0.002 mm in equivalent diameter. (*ii*) A soil textural class. (*iii*) A time grained soil that has a high plasticity index in relation to the liquid limits (engineering). (*iv*) A Specific mineral structure. (v) Soil material that contains 40 per cent more clay less than 45 per cent sand, and less than 40 per cent silt.

Clay films: Coatings of clay on the clay surfaces of soil pads and mineral grains and in soil pores. (Also called clay skins, clay flows illuviation cutans, argillans or tonhautchen.)

Clay pan: A dense layer in the subsoil having a much higher clay content than the overlying material, from which it is separated by a sharply defined boundary; formed by downward movement of clay or by synthesis of clay in place during soil formation. Clay pans are usually hard when dry, and plastic and sticky when wet. Also, they usually impede the movement of water and air, and the growth of plant roots.

Clean cultivation: Periodic soil tillage to eliminate all vegetation other than the crop being grown.

Cleavage: The smoothness of regular breaking surface of minerals, where "perfect" is very smooth and flat; "conchoidal" is convex and not smooth. Cleavage helps in mineral identification.

Cliestogamy: (*a*) Flowers do not open at all, *i.e.*, there is no anthesis. (*b*) A built-in breeding mechanism where flowers remain closed at the time of pollination. It favours self-pollination.

Climacteric fruit: Fruits in which the respiration rate is minimum at maturity and remains rather constant even after harvest which gradually increases at the beginning of ripening followed by a sharp rise to a peak (climacteric peak) and then slowly declines (post-climacteric stage).

Climate: The aggregate of weather conditions over a long period of time.

Climatic index: A number which condenses climatic data into a simplified expression (Transeau's ratio P/E is climatic index).

Climatology: The study of weather patterns over time and space. It concerns with the integration of day-to-day weather over a period of time.

Climax: The most advanced succession community of plants capable of development under, and in dynamic equilibrium with, the prevailing environment.

Climax vegetation: The final stage of plant succession in which the plant community reaches a steady state in balance with the soil, climate and fauna of a given area.

Clod: A compact, coherent mass of soil ranging in size from 5 to 10 mm to as much as 200 to 250 mm produced artificially, usually by the activity of man in ploughing, digging etc. especially when these operations are preferred on soils that are either too wet or too dry for normal tillage operations.

Clonal multiplication: Multiplication through asexual reproduction.

Clonal selection: A breeding method of asexually propagated plants based on the selection of superior clones from a wide range of clones on visual observation of morphological traits.

Clone: All the individuals obtained from a single individual through asexual reproduction (*i.e.*, produced by mitotic cell division).

Cloning: Multiplication of a clone by means of vegetative propagation.

Closed ended questions: Closed ended questions are questions having limited number of responses.

Closed formula mixed fertilisers: The fertiliser grade is disclosed on each bag of such fertiliser mixture, but the ingredients or straight fertiliser used in formulating the mixture are not disclosed. In India, fertiliser mixture sold to the cultivators is usually to the closed formula type.

Cloud seeding: Injecting the clouds with seeding agent like dry ice, silver iodide, sodium chloride from an aircraft or using a ground generator for producing artificial rain.

Clustering/Jhoomka: A disorder in mango characterized by a cluster of fruitlets at the tip of the panicles due to poor pollination and fertilization.

Coarse texture: Included the sand, loamy sands, and sandy loams except the very fine sandy loam textural classes. Sometimes sub-divided into sandy and moderately coarse textured.

Coated fertiliser: Fertiliser, the granules of which are covered with a thin layer of different material in order to improve the behaviour and/or modify the characteristics of the fertiliser.

Cobblestone: Rounded rock and mineral particles of size between 8 and 25 cm diameter.

Cobbly: Contain appreciable quantities of cobblestones. The term angular cobbly is used when the fragments are less rounded.

Codominance: Both the alleles of a gene express themselves in the heterozygotes.

Codon: A group of three adjacent nucleotides of mRNA which code for one amino acid; *nonsense codons* specify no amino acid and function in polypeptide chain termination.

Coefficient of Coincidence: In case of linkage; the ratio of observed frequency of double crossovers to their expected frequency.

Coefficient of Interference: One minus the value of coefficient of coincidence; in case of linkage.

Coefficient of Variation: The ratio of standard deviation of sample to the mean of that sample; usually expressed in per cent.

Cognitive learning: Learning that deals with facts, procedures or concepts.

Cohesion: Holding, together force holding a solid or liquid together because of attraction between like molecules; cohesion decrease with rise in temperature.

Colchicine: An alkaloid derived from autumn crocus; it prevents the organisation of spindle apparatus and, thereby, induces chromosome doubling.

Cold fog: The cold fog occurs during winter in cool, moist, temperate latitude when extremely low temperatures (up to –30°C) occur, resulting in ice fog of small particles.

Cold frame: An enclosed (glass or plastic) unheated frame used as a propagation unit and for growing or protecting young plants in early spring in a temperate climate.

Cold hardiness: A process by which a plant acquires hardiness to withstand cold conditions due to continuous exposure to low temperatures.

Cold storage: An insulated storage utilizing mechanical refrigeration to maintain a stable cold temperature for long-term storage of agricultural products.

Cold wave: The cold wave is defined as appreciable fall in minimum temperature (cold wave –3 to –4°C).

Coliform: A general term for a group of bacteria that inhabit the intestinal tract of human and other animals. Their presence in water constitutes presumptive evidence for fecal contamination. Includes all aerobic and facultative anaerobic, gram-negative rods that are non-spore forming and that ferment lactose with gas formation. *Escherichia coli* are important menders.

Colinarity: The sequences of mutant alleles in the linkage map of a gene being identical with the sequences of altered amino acids generated by these alleles in the polypeptide specified by the gene in question.

Collar: The part of the plants where the roots meets the stems.

Colloid: A fine particle usually 10^{-6} to 10^{-4} mm in diameter which carries an electric charge. A wet mass of colloidal particles is glue-like in consistency.

Colluvium: A general term applied to deposit on a slope or at the foot of a slope of cliff that were moved there chiefly by gravity. Talus an cliff debris are included in such deposits.

Colour Blindness: Inability of an individual to distinguish between two colours, *e.g.*, red-green colour blindness; usually, sex-linked.

Colour sport: A clone resulting from a mutation in the bud which leads to increased amount of anthocyanins in the outer cell layers of the fruit skin and the variation in colour of flower or fruit of a established cultivar.

Combination breeding: It involves the transfer of one or few characters, usually governed by oligogenes, from one variety into another.

Combined tillage: Operations simultaneously utilizing two or more different types of tillage tolls implements (subsoil's r-lister, lister-planter or plough-planter combinations) to simplify control or reduce the number of operations over a field.

Combining ability: Ability of a strain to produce superior progeny upon hybridization with other strains.

Cometabolism: Transformation of substrate by a microorganism without deriving energy, carbon, or nutrients from the substrate. The organism is able to transform the substrate into intermediate degradation products but fails to multiply at its expense.

Command area: Area in the vicinity of a water source that gets the benefits of irrigation from the source. The water source could be a tank, well, pond or a reservoir.

Commercial cash crop production: It involves growing crop primarily for sale not for home consumption.

Commercial cash energy or non-renewable source of energy: Energy from fossil sources such as coal and petroleum.

Commercial fertiliser: It has been indicated that three elements most commonly deficient for crop production are nitrogen, phosphorous, and potassium. Therefore, a complete commercial fertilizer contains all three of these elements. The analysis (grade) of the fertiliser is required by state law to be shown on the bag and declares the per cent by weight of nitrogen, phosphorous, and potassium in that order. The latter two may be expressed as oxides (P_2O_5 and K_2O). The analysis will appear as number such as 24-10-8 on the bag.

Common land: Land collectively owned and managed by a defined group of users, ideally governed by a common property regime (system of rights and duties) which prevents over-exploitation: to be differentiated from open-access, which has no barriers to its use.

Common property resources: The term used to describe land owned by the entire community with no individual having the right to ownership. Such lands are used as common grazing grounds and serve to meet the fodder and fuel requirements of the community. Upgradation of these lands can also be taken up by programmers under social forestry, afforestation, avenue plantation and block plantation.

Community forestry: It is the forest of the community by the community and for the community. Such forestry is managed for the interest of local community or village and then run almost exclusively by them.

Compact disc-read only memory: The CD-ROM is one of the powerful and interactive form of computer based technologies. The training materials in the computer are provided in the form of text, graphics, animation and sound.

Compatible: Two compounds or products are said to be compatible when they can be mixed without affecting each others performance.

Competence: The ability of a bacterial cell to take-up exogenous DNA and, as a result, to become transformed.

Competition: Relative success of individuals in utilising critical parts of the local environment. Also termed as interference.

Competitive ability: The ability of genotype to survive in a mixture of several genotypes.

Competitive antagonism: The type of antagonism that occurs when an antagonist acts reversibly at the same site as the herbicide.

Complementarily: A state in which one element, in combination with one or more other elements, completes the whole.

Complementary genes: Two (or more) dominant genes capable of producing a character only when both (or all) of them are present together in the dominant state.

Complementation: The F_1 obtained by crossing two mutant strains has the wild type phenotype.

Complementation test: A test to determine, on the basis of complementation, if two mutant alleles are located in the same gene; if there is complementation, the mutants are located in different genes, otherwise they are located in the same gene; cases of intragenic complementation known, especially in eukaryotes.

Complex fertilisers: The commercial fertiliser containing atleast two or more of the primary essential nutrients, they are designed as incomplete complex fertilisers, while those containing all three primary nutrients are called as complex fertilisers.

Composite cross: A hybrid population generated from a large number of pure lines either by composting equal seeds from the F_1 or F_2 generations from all possible crosses among them, by ordering them into a complex cross, or allowing them to cross with a male sterile line and harvesting the seeds from only the male sterile plants in each generation.

Composite varieties: Varieties produced by open-pollination among a number of outstanding strains usually not tested for combining ability with each other.

Compost: Organic, residues or a mixture of organic residues and soil which have been mixed, piled and moistened with or without addition of fertilizer and lime, and generally allowed to undergo thermophilic decomposition until the original organic materials have been substantially altered or decomposed. The final product can be easily worked into the soil, or can be used in or as a potting mix. Sometimes called artificial manure or synthetic manure.

Composting: A biological process in which micro-organisms of both types, namely, aerobic and anaerobic, decompose the organic matter and lower the carbon nitrogen ratio of the refuse. The final product of composting is well-rotted manure known as compost.

Compound fertiliser: Fertiliser having a declarable content of at least two of the nutrients nitrogen, phosphorous and potassium obtained chemically or by blending or both.

Compressibility: The property of a soil pertaining to its susceptibility to decrease in bulk volume when subjected to a load.

Computer based training: Computer based training is an interactive training where the learner using keyboard or mouse can interact with the programme and respond to the questions or activities.

Concentrated flow: The flowing of a rather large accumulated body of water over a relative narrow coarse. It often causes serious erosion and gullying.

Concentration: The amount of active ingredient in a given volume of diluent or given weight of dry material.

Concordance: The two members of pair, *e.g.,* identical twins, showing a similar phenotypic expression for a trait.

Concrete experience: The stage of learning cycle that emphasizes learning from specific experience through direct involvement.

Concretion (soil): Hardened, local concentration of certain chemical compounds, such as calcium carbonate or iron oxides, in the form of indurate grains or nodules of various sizes, shapes and colours.

Condensation: The process of formation of liquid from its vapour.

Cone penetrometer: A instrument in the form of a cylindrical rod with a cone-shaped tip designed for penetrating soil and for measuring the end bearing component of penetrating resistance to penetration developed by the cone equals the vertical force applied to the cone divided by its horizontal projected area.

Confounding: When the number of treatment combinations are large the plot-to-plot variation within blocks on which is based the estimate of experimental error would thus increase and lower the precision of the experiment. The principles of confounding is to divide the replication into two or more compact blocks of suitable sizes to ensure that the estimate of unconfined effect are obtained with increased precession by removal of the effect of possible heterogeneity with a large replicate.

Conjugation: The union between individuals in unicellular organisms; individuals of different mating types either fuse completely (*Chlamydomonas*), exchange nuclei (*Paramecium*), or may show one-way-transfer of the genetic material (bacteria, *e.g., E. coli*).

Conscious competence: The stage in the learning process where the trainees gets exposure with the advanced knowledge on technology that can be used to improve his work efficiency but still needing a conscious effort to achieve the required standard.

Conscious incompetence: The stage in the learning process where the trainees slowly starts become aware about the new developments around him, develops consciousness on the advances in his subject and form interest towards it.

Conservation: The protection, improvement and use of natural resources according to principles that will ensure their highest economic social, and psychological benefits to people in perpetuity.

Conservation cropping: A way of farming that aims to maximize the protection against erosion that can be achieved through soil and crop management for sustained farm productivity.

Conservation tillage: A tillage designed to maintain roughness of a field surface and leave most of the previous crop residues on the surface while providing a suitable seedbed and weed control for the crop. The roughness reduce water and run off and water wind erosion.

Consistence (soil): (*i*) The resistance of material to deformation or rupture. (*ii*) The degree cohesion or adhesion of the soil mass.

Constant-charge surface: A mineral surface carrying a net electrical charge whose magnitude depends only on the structure and chemical composition of the minerals itself. Constant charge surfaces usually arise from isomorphous substitution in mineral structure.

Constant-potential surface: A solid surface carrying a net electrical charge which may be positive, negative, or zero, depending on the activity of one or more species of ion (called a potential-determining ion) in a solution phase contacting the surface. For minerals common in soils, the potential determining ion is usually H^+ or OH^-, but any ion that forms a complex with the surface may be potential determine.

Constitutive enzyme: An enzyme which is produced at a constant rate wither it is needed by the organism or not.

Consumptive irrigation requirement: It is the depth of irrigation water exclusive of precipitation, stored soil moistures or ground water that is required to meet evapo-transpiration demand of the crop.

Consumptive use: Quantity of surface and ground water absorbed by crops and transpired or used directly in the building of plant tissues, together with that evaporated from the cropped area, expressed in units of volume per unit area.

Contact herbicide: A chemical that kills primarily by contact with plant tissue rather than as a result of translocation.

Contact placement (fertiliser): The drilling of seed and fertiliser together while sowing.

Contiguous control: A control or check variety planted in an adjacent plot; design for the valuation of a large number of lines/progenies in which the control variety is planned in every 5th or 7th row/plot (augmented design), and the comparison of the line/progeny means are made with the control.

Contingency cropping: Contingency cropping is growing of crops in aberrant situations like drought and floods. It aims at partial mitigation of misery by producing some food, feed and fodder to encounter emergency conditions.

Contingency planning: Planning to face unforeseen problems.

Continuity: The capability to conserve the natural, financial and human resources of a farm system so as to ensure its persistence.

Continuous variation: Variation not separable into distinct classes; governed by ploygenes; .markedly affected by the environment; in quantitative traits; analysis based on measurement data.

Contour: A contour is an imaginary line joining different points of the same elevation. The art of tracing a contour along the line of equal elevation is known as contouring. Contouring presents the layout of the land on the basis of which appropriate methods of land treatment can be employed.

Contour border irrigation: Border strip irrigation practiced along the contour.

Contour bunding: The construction of small bund across slope of the land on a contour so that the long slope is cut into a series of small ones and each contour bund act as barrier to the flow of water.

Contour bunds: Created exactly along contours to impound runoff water. The impounded water gradually infiltrates into the soil for use by crops. These bunds are used in areas with less than 600 mm rainfall or with highly permeable soils.

Contour cropping: The cultivation of crops along contours of a slope.

Contour Cultivation: A method of farming along the contours wherein all the operations including sowing, planting and ploughing are carried out across the slope, along the contour lines. The numerous ridges and furrows that are created as a result of this process help in harnessing an adequate amount of runoff water while reducing soil erosion to a considerable extent.

Contour farming: Growing crop between contour lines stabilised by, *e.g.*, earth bunds, stone ridges or hedges which conserve both soil and water; the system may also include a livestock component.

Contour intervals: A contour interval is the difference in elevation between any two adjacent contours on a map. The vertical distance between any two consecutive contours is called the vertical interval while the horizontal distance is called the horizontal interval. Vertical and horizontal intervals are fixed on the basis of the slope percentage and the activity (for example: crop management, horticulture or forestry) taken up in the area. Recommendations for fixing these intervals are usually given by the department of agriculture of the respective states.

Contour irrigation: A method of irrigation where the water is applied to a field or orchard, which has been divided into strips along the contours.

Contour plantation: Planting row to crops along the contours.

Contour ploughing: Ploughing done along the contours.

Contour strip cropping: The cultivation practices involving growing of a soil exposing and erosion permitting crop in strip of suitable widths alternating with strips of soil protecting and erosion resisting crops, along the contours.

Contour terrace: A low terrace formed along the contour on steep mountains slopes with inadequate natural cover to prevent erosion until natural cover becomes established.

Control: In plant breeding; a known variety used for comparing the performance of newly developed strains their evaluation.

Controlled atmospheric (CA) storage: Storage of commodities under gaseous atmosphere, like gas storage, where the composition of atmosphere is controlled accurately. The modified atmosphere and controlled atmosphere differ only in degree of control and CA is more exact.

Convective rain: Rainfall which is vertical motion of an ascending mass of air which is warmer than its environment; the horizontal dimension of such an air mass is generally of the order of 15 kilometres or less and forms a typical cumulonimbus cloud.

Conventional tillage: The combined primary secondary tillage operations normally performed in preparing a seedbed.

Convergent cross: A cross involving more than two parents (*Syn.*, complex cross).

Convergent questions: The convergent questions are deductive in nature focusing on lower level of thinking.

Converges: Trainees with convergent skill and rely more on abstract conceptualization and active experimentation.

Cool chain: It refers to the practice in which the produce is maintained in low temperature environment from field to consumer in a fresh form. It implies to the construction of cold storage in the production zones, refrigerated trucks or rail, transporting refrigerated ships and development of cold storage at retail shops.

Co-operative better farming: A type of co-operative farming where the land is not pooled and the cultivation is carried on by each farmer separately. A member is free from his own way of farming except in respect of the purpose for which he has joined the society, *e.g.*, for irrigation, purchase of seed or marketing of produce etc.

Co-operative collective farming: A type of co-operative farming wherein the land is owned by the society and cultivation is carried out jointly. The member work on the land under the direction of a managing committee. The profits are paid to the members in proportion to the work and capital contributed by each member. The right share of individual member in the land is not recognized.

Co-operative farming: Co-operative farming means a system under which all agricultural operations or part of them carried out by the farmers on a voluntary basis, each farmer retaining right of his own land. The farmer pools their land, labour and capital. The land is treated as one unit and cultivated jointly under the direction of an elected management. A part of profit is distributed in proportion to the land contributed by each farmer.

Co-operative joint farming: In this farming, the land of members is pooled for joint cultivation. The ownership of each member over his own land is recognized by payment of dividend in proportion of the value of his land. The members work under the direction of the managing committee and each member receives for his daily labour.

Co-operative tenant farming: A type of co-operative farming wherein the land is held by the society and not by the members independently. The land is then dividend into plots, which are leased out to the members. The society arranges for agricultural requirements, *e.g.*, credit, seeds, manure are marketing of the produce etc. Each member is responsible to the society for payment of the rent of his plots. He is at liberty to dispose of his produce in such a manner, as he likes.

Co-ordinated varietal trail (CVT): Multilocation trails (usually for two years) conducted under the respective All India Co-ordinated Crop Improvement Projects for the evaluation of the performance of newly developed strains; in many corps, one year of IET precedes CVT (*Syn.*, UR, Uniform regional Training).

Copper sulphate: It is the most common salt used in fertiliser, commercial grade of copper sulphate supplies 25-35 per cent copper (Cu).

Coppice: A form of woodland management in trees is cut back regularly to encourage growth of numerous adventitious shoots from the base and the result thicket is turned a coppice or copse.

Coppice farming: The word 'coppice' has been defined as the ability to regenerate by shoot or root suckers, or a forest so established. Thus coppice farming is a practice of incentive exploitation of land capability by regular harvesting of such trees which produce numerous shoots from the stump after cutting. Coppice sprouts usually grow vigorously because they are served roots big enough to feed the former tree.

Coppice mound: A small mound of stabilized soil material around desert shrubs.

Copping system: The physical and temporal arrangement of crop specials usually described within a given area.

Coprogenic material: Remains of fish excreta and similar materials that occur in some organic soils.

Cordon: Three or more stems

Core breakdown: A storage disorder of pear which appears as a brown, watery collapse of cortex tissue, beginning near the core of the fruit and often accompanied by darkening of the main vascular elements between the stem and core. This is greatly influenced by handling and storage.

Core browning: A storage disorder of apple characterized by diffuse browning of tissues in the core area next to the carples, occurring after 3-4 months of cold storage (–0.5 to 3.5°C) and becoming more extensive the warm temperature after storage.

Core spot of pear: Bumpy, uneven appearance of the surface as fruit approach maturity, with large masses of brown or grayish necrotic cells underlying individual spot.

Cork spot (drought sport): A physiological disorder of pear fruits similar to bitter pit of apple.

Cork spot of apple: Pitting of the flesh like bitter pit due to deficiency of boron which develop only on the tree.

Corky core: A disorder of fruit in which the tissue around the core turn brown and become more or less corky resulting a bitter taste.

Correlated Response to Selection: Change in one or more quantitative characters due to selection for another character.

Cortex: The tissue region between the epidermis and vascular system of a plant.

Cost of cultivation: Total expenditure involved in raising a crop, including rental value of the land.

Cotton bale: A pressed bale is ginning and pressing factory, which weighs about 178 Kg of lint.

Cotton boll: The fruit of cotton plant.

Cotton seed cake: A by-product of the cotton seed crushing industry. When it is undercorticated, it is non-edible containing 3.9 per cent nitrogen, 1.8 per cent phospheric acid and 1.6 per cent potash. When decorticated, cotton seed cake becomes edible oilcake containing higher percentage of plant nutrients. Decorticated cotton seed cake contains 6.44 per cent nitrogen, 2.9 per cent phosphoric acid and 2.2 per cent potash.

Cotyledon: The first leaf or pair of leaves (depending on whether the plant is monocotyledonous or dicotyledonous) of the embryo of seed plants.

Coulter: A disk used in tillage to cut surface or open a silt to facilitate the passage of following chisels or knives in reduced tillage.

Coulure: Failure of grape blossoms to set, resulting in premature drop.

Coupling phase: In linkage, the dominant alleles of two or more genes present in the same chromosome and, hence; linked together, contributed by the same parent.

Course contents: The course contents are topic, formulated keeping into consideration of learning domains such as knowledge, skill and attitudes, to be delivered through the training programme.

Cover crops: Crops are grown primarily to cover the soil and to reduce the loss moisture due to leaching and erosion by wind and water. Annual crop sown to create a favourable soil micro-climate, decrease evaporation and protect soil from erosion. Cover crops also produce biomass, which can be used for soil fertility management.

Cracking: A disorder where fruit surface cracks mainly due to heavy irrigation or rain after long dry spell. This may occur due to varietal characters and micro nutrient deficiencies; generally found in tomato, litchi, cherry, apple, pomegranate etc.

Creasing of citrus: Primarily a pre-harvest disorder where grooves or furrows are irregularly distributed on the fruit surface.

Creep (soils): Slow mass movement of soil and soil material relatively steep slopes, primarily under the influence of gravity but facilitated by saturation with water and by alternate freezing and thawing.

Crescent bunds: A semi-circular barrier built on the downstream side of the slope around a plant or a tree. It collects and stores water thus increasing the infiltration rate.

Crest: The slope component that is commonly at the top of an erosional ridge, hill, mountain, and so on.

Critical period or critical stage crops: The period or the stage development in the life cycle of a crop or a plant when it is most sensitive to the deficiency of a production factor and is most responsive to the correction of the deficiency.

Critical period (Water): It is the period during which a crop is affected severely due to moisture stress and the loss cannot be compensated by adequate moisture supply on other periods or stages.

Critical photo-period: It is the day length value separating inductive range of photo-period from non-inductive range. For example, the critical photo-period for flowering in Xanthium is 15 hrs.

Crook stage: The stage after bean seedlings has broken through the soil but before the stem has become erect.

Crop: Annual or perennial plants cultivated to yield products desired for human consumption or processing, *e.g.,* grain, vegetables (edible root, stems or leaves), flowers, fruit, fibre, fuel.

Crop cafeteria: Crop cafeteria is the demonstration of identified efficient crops/varieties is an agro-metrological region/zone offering an opportunity to the farmer to choose a suitable crop or crops. This approach is also applicable in case of production technology where a number of them are demonstrated and the farmer has a choice to choose a suitable one.

Crop calendar: A list of the standard crops of the region in the form of a calendar giving the dates of sowing and various operations including harvesting to be done during the crop season in years of normal weather.

Crop canopy: It is the structure of aerial vegetative parts with special reference to the size, orientation, depth, density and arrangement of leaves influencing the penetration and interception of radiant energy.

Crop competition: Two plants are in competition with each other when the growth of either one or both of them is reduced or their form modified as compared with their growth or form in isolation.

Crop ecology: The branch of 'plant ecology' which deals specially with the study of the inter-relation amongst crop plants and environment including measurement practices.

Crop efficiency zone: A geographical area characterized by high speed of a crop combined with high yields consistent with minimum variations from year to year.

Crop environment: A crop environment may be regarded as having components, the gross environment which takes into account the general soil and atmospheric conditions outside the crop, and also the changes of those general conditions caused by the plant.

Crop evapo-transpiration: Rate of evapo-transpiration of a disease free crop growing on a large field under optional soil conditions, including sufficient water and fertilizer and achieving full production potential of the crop under the given growing environment: includes water loss through transpiration by the vegetation and evaporation from the soil surface and the wet leaves, mm/day.

Crop modelling and simulation: Crop modelling and simulation is a simplified representation of a more complex reality in crop modelling system. Simulation is the operation or manipulation of a model system to estimate the consequence of selected strategies or combination of input plants.

Crop planning: The selection of a particular variety of crop based on the following conditions: Climate, Soil type, Minimum support price, Marketing facilities, Local requirements.

Crop production: Crop production is concerned with the exploitation of plant morphological (or structural) and plants physiological (or functional) response within a soil and atmosphere environment to produce a high yield per unit area of land.

Crop production strategy: It refers to a sib-set of the farm production strategy that involves only as its components. It is described by the spatial and temporal arrangement of crops to be raised in a parcel of lands (cropping pattern) and the procedure for growing each component crop. A crop production strategy can be either one mono-cropping or multi-cropping.

Crop productivity: Economic yield or production of plant product of economic importance, expressed in standard units per unit area.

Crop protection: Measures taken to protect the plant from weeds, pests and diseases. They include the use of Cultural practices, Pesticides, An effective combination of cultural practices and pesticide use. The measures could be both preventive and prophylactic.

Crop residue: The portion a plant or crop left in the field after harvest, or that part of the crop that is not used domestically or sold-commercially.

Crop residue management: Use of the non-commercial portion of the plant or crop for protection or improvement of the soil.

Crop rotation (sequential cropping): The growing of different crops on a piece of land in a pre-planned succession.

Crop season: The most favourable condition to get better yields.

Crop system: Composes all components required for the production of a particular crop and the inter-relationship between them and environment.

Crop technology: The theoretical and practical approaches to crop production in which concepts of applied science and advanced techniques are related to the financial consequence of their use.

Crop yield indexes: A measure of comparison of the yield of all crops on a given farm with the average yield of these crops in the locality. The relationship is expressed in percentage.

Cropping index: The number of crops grown per annum on a given area of land times 100.

Cropping intensity: It is the ratio between total cropped area and actual net cultivated area expressed in percentage. It does not take into consideration the length of growing period for various crops.

Cropping pattern: Yearly sequence and spatial arrangement of crops and fallow on a given area.

Cropping systems: Cropping systems may be defined as the cropping pattern used on a farm keeping in mind the farm resources, other farm enterprises and the technology available.

The two main cropping patterns are—Monocropping and Multicropping.

Cross: Mating of two or more individuals or strains having different genotypes. Also the product of such mating.

Cross resistance: In the context of weeds, it is an expression of mechanism that confers plants the ability to resist or withstand herbicides from different classes.

Crossing over: An exchange of homologous segments between the nonsister chromatids of homologous chromosomes leading to recombination between linked genes; inferred on the basis of recombination between linked genes; demonstrated cytologically; occurs during pachytene.

Crossover unit: The distance between two linked genes which permits 1 per cent recombination between them (*Syn.,* may unit, Morgan).

Cross-pollination: *See*, allogamy

Crown: The point where stem and root join in a seed plant.

Crown grafting: A method of grafting in which a mature dormant wood of the previous season growth is grafted on the stock in the crown region, i.e. the transition region between the stem and root.

Crushing strength: The force required to crush a mass of dry soil or, the resistance of the dry soil mass to crushing. Expressed in units of force per cent area (pressure).

Crust: A soil-surface layer, ranging in thickness from a few millimetres, that is much compact, hard, and brittle when dry than the material immediately beneath it.

Cryic: An annual soil temperature regime (0-8°C) in the quite cool environment.

Cryogenic soil: Soil that has formed under the influence of cold soil temperatures.

Crystal: A regular arrangement of atom in space.

Crystal structure: The orderly arrangement of atoms in a crystalline material.

Crystalline rocks: crystalline rocks include ancient granites, gabbros, basalts, and associated metamorphic rocks that make up the foundation of the continents. These rocks are buried out of sight for the most part, often to great depths under beds of sandstone and shale.

Culm: The jointed stem of grass or bamboo which is usually hollow.

Cultigen: Plant or group of plants knows only in cultivation which presumed to have originated under domestication.

Cultivable or culturable command area: Gross command area minus uncultivated area. Command area is that which falls within irrigable area of project.

Cultivar: An assemblage of cultivated plants which is clearly distinguished by any character and which when reproduced, sexually or asexually, retains raising crops.

Cultivation: A tillage operation used in preparing land for seeding or transplanting or later for weed control and for loosening the soil.

Culture fisheries: Cultured in constructed ponds and other enclosures, in which fish are fertilized and provided with supplementary feed to achieve economic returns.

Cumulative run off: The total volume of run off in a specified period of times.

Cuticle: A thin, continuous, non-cellular, lipoidal membrane (varnish-like layer) covering the entire plant surface (shoot), formed by oxidation of plant oils.

Cutin: The chief structural component of cuticle, composed of polymerized long-chain fatty acids and alcohols, protecting the photosynthetic mechanisms of plant cell.

Cuttings: Various plants vegetative materials used in propagation.

Cyanobacteria: New name for N_2 fixation microbes, which were previously called blue-green algae.

Cybrids: Somatic hybrids obtained by fusion of complete protoplast of one species with such protoplast with another species which either have no nucleus or have no inactive nucleus, *e.g.*, *protoplast of species A + cytoplast of species B = cybrid.*

Cyclic salt: Salt derived from the sea or salt lakes deposited on the landscape from wind or rainfall.

Cyclonic motion: If the motion of air is in the counter clockwise direction round the centre in Northern Hemisphere is called cyclonic motion.

Cyme: An inflorescence whose single terminal flower opens first, followed by those of secondary and tertiary axes (as in apple).

Cytochimera: It refers to chromosomal chimera, i.e. different chromosome numbers in layers in the same part of the plant, which is similar but not necessary identical to a mixoploid.

Cytogenetics: A branch of biology devoted to the study of chromosomes and their implications in genetics.

Cytokinesis: The division of cytoplasm into two or more parts; usually follows caryokinesis or nuclear division.

Cytology: A study of the structural and functional organisation of the cell.

Cytoplasm: The entire content of a cell other than the nucleus includes ER, plastids, mitochondria, microtubules etc., and the hyaloplasm.

Cytoplasmic inheritance: Inheritance due to the genes located in the cytoplasm (*plasma genes*), *e.g.*, mitochondria and plastids; typical uniparental transmission and the lack of segregation; in higher plants and animals, transmission through the female parent. (*Syn.*, extranuclear, extrachromosomal and maternal inheritance).

Cytoplasmic male sterility: Cytoplasmic male sterility for which a restorer gene is available/known.

Cytoplast: Protoplast containing either no nucleus or an inactive nucleus.

D

Dairy husbandry: The care, breeding, feeding and milking or dairy cattle and production and sale of milk.

Dam: A small barrier constructed across a water course in order to reduce the velocity. Dams helps in minimising the channel scour, inducing the deposition of eroded material and encouraging percolation. There are several kinds of dams. In the upper reaches of the terrain, due to the relatively low water volume—live check dams, loose boulder structures of earthen checks are recommended. In the middle reaches of the terrain, where the gullies are well formed and the water volume is higher—brushwood structures, loose boulder structures or earthen checks are recommended. In the lower reaches of the terrain, where the drainage courses run deep and the volume of water is at its highest, dams are made to suit the specifications of measurement. They include earthen dams, small masonary structures and rock filled dams.

Dam, brushwood: A brushwood dam is made up of cut and broken branches, dead roots, twigs and other locally available materials. Those dams are used for the control of gullies, and are recommended in places where it is not possible to build a firm foundation for rock filled dams or in places where stones are. The centre of brushwood dam is made lower than sides, to allow water to flow over the dam. Brushwood dams slow the runoff and reduce soil erosion.

Dam, check: A check dam is usually built withcheap materials of temporary as well as permanent value. The materials used could be rubble loose rock, sand bags etc. These structures do not require the use of expensive material and the construction is also not done to precision. Check dams are usually found in the upper reaches of the terrain or at the gully heads.

Dam, earthen: A mechanical barrier made of soil reinforced at times, with stone revetments. These dams are stronger than brushwood dams. They are usually found in the middle reaches of the terrain, in black soil areas.

Dam, masonary: Masonary dams are permanent structures made of brick, concrete or cement or a combination of all these materials. They are used to control a relatively large flow of water and are built to specific measurements.

Dam, rock filled: Rock filled dams are found in areas where there is plenty of rock available. They are used to control gullies of moderate slope with small to medium sized drainage areas. These dams are usually found in the middle and the lower reaches of the terrain. A well constructed rockfilled dam can be made more durable than brushdam or a check.

Dao: A local utility implement or tool used by the Nagas for cutting and chopping.

Dapog rice-nursery: Rice seedlings are raised on well-prepared soil (wet bed) either in field or in nursery boxes, covered under banana leaves or plastics with bamboo pegs. Dopag seedlings are ready for transplanting in 9-14 days after the seed is sown. It is labour and time saving method with ease of transport.

Day length: Day length is the interval between sunrise and sun set.

Day neutral plants: The plant that are independent of day length and can bloom under conditions of either long or short days.

Dealklisation (solodisation): The leaching of sodium ions and salts from the soil profile.

Deblossoming: Removal of flowers from the plant to reduce the crop load, done either by withholding water or by spraying of chemicals or mechanical means.

Decalcification: Removal of calcium carbonate or calcium ions from the soil by leaching.

Deciduous: Plants or tree that shed their leaves at a particular season, usually winter.

Decoction: A preparation made by boiling a substance in water or other liquid.

Decomposition: The breakdown of minerals and organic materials.

Deep seepage: The portion of the run off which escape from reservoir through the underlying earth of rock strata below any possible intercepting cut-out constructed at the dam.

Deep soil: A soil having a solum depth of more than 100 cm.

Deepening: In synoptic meteorology, 'deepening' of a depression signifies a decrease of pressure at a centre of the system with time.

Deep-water rice: Rice grown with more than 51 cm 5-6 metres of standing water.

Deficiency: The loss of a segment from a chromosome; deficiency heterozygotes are hemizygous for the genes located in the deleted segment; many deficiencies produce genetic effects similar to gene mutations (*Syn.*, deletion).

Defoliant: A chemical that causes the foliage to abscize from a plant.

Degenerate code: One amino acid is specified by more than one codon.

Degree day: It is measurement of the departure of the daily temperature above the minimum threshold temperature for a crop or plant.

Degree of freedom: The number of degree of freedom is the number of independent comparison that can be made within the body of data. In χ^2 test, the number of classes to which an arbitrary number of observations can be assigned to yield the given total; ordinarily it is one less than the number of classes in the data.

Degreening: The process of decomposing the green pigment in fruits by applying ethylene (1000-2000 ppm) or similar metabolic inducers to give a fruit to its characteristic colour as preferred by consumers, generally followed in citrus fruits but also practised banana and mango.

Dehiscence: The process of opening of an anther or a seed pod.

Delineation: The process of demarcation of the boundary of a watershed into a hydrological and physiographical entity, encircled by a distinct ridge line. Delineation is essential for the planning and implementation of different activities under a watershed programme. It takes into account the physical and socio-economic resources of the area under the watershed programme.

Delinter: A machine used to remove residual lint (fuze) from cotton seed.

Demonstration: Demonstration is a presentation which illustrates a task, procedure, or the use of an equipment in a step-by-step fashion so that trainees can observe and understand the procedure, principles or phenomenon in an interesting and convincing way.

Denaturation: The loss of natural or native configuration of a macromolecule; generally, accompanied with a loss in the biological activity of the molecule; a separation of the two strands of DNA double-helix producing single-stranded DNA molecules.

Denitrification: The biochemical reduction of nitrate or nitrite to gaseous nitrogen in the soil either as molecular nitrogen or as an oxide of nitrogen usually carried out by denitrifying bacteria, which result in escape of nitrogen to the atmosphere.

Departure: The amount positive and negative, by which the value of a meteorology element departs from the normal, average or expected value of a place and time. It is also termed as anomaly.

Depletion phase (irrigation): Portion of the total irrigation time between inflow shut off and beginning of recession at upper end of the field.

Dermatitis: Inflammation of the skin.

Desalinisation: Removal of salt from saline soil, usually by leaching.

Desert: Places where rainfall is often less than 250 mm per year with extreme temperature fluctuations. Most of the vegetation is restricted to living only during the very short rainy season.

Desert soils: A large part of western Rajasthan, Haryana and Punjab, lying between the Indus river and the Aravalli range is affected by desert conditions of geologically recent origin. This part is covered under a mantle of sand combined with the arid climate, resulting in poor soil development. The predominant component of desert sand is quartz in well-rounded grains, but feldspar and horneblende grains also occur with a fair proportion of calcareous grains. Some of these soils have high percentages of soluble salts, possess high *p*H, a varying percentage of calcium carbonate and, are poor in organic matter.

Desertification: Process of continued decline in the biological productivity of arid/semi-arid land, resulting in skeletal soil that is difficult to revitalise: refers here also to land degradation, *i.e.*, reduction in the capability of land to satisfy a particular use.

Desiccant: A chemical that promotes or accelerates dehydration of plant tissue, causing drying; it also lowers moisture content of seeds to facilitate *seed* harvest.

Design: A method of arranging sample or experimental plots to minimise the effects of uncontrolled variations in fertility and other natural factors and to make possible estimate of such effect in relation to those due to variations in treatment.

Desorption: Movement of particles of chemical substances (*e.g.*, herbicides) from the soil surface into the soil solution; the reverse of adsorption.

Desuckering: The act or processing of removing suckers.

Detasseling: Removal of young rassels (pollen producing organ) in maize plants before pollen is released so as to render the plants fully female especially in hybrid maize seed production.

Detergent: A chemical having the ability to remove oil. These may be used as surfactants in herbicidal sprays which remove way substances found on many of the weed leaves.

Determinate: (*a*) When the terminal flower opens first and the prolongation of the axis is thereby arrested or develop cessation of growth at the apex or in the main axis. (*b*) With the terminal (centre) flower of an inflorescence opening first, thus stopping terminal growth. The lateral flowers open somewhat later.

Determination: The process due to which an embryonic cell becomes committed to differentiate only into a specific adult organ.

Development: In synoptic meteorology, the intensification of circulation, cyclonic or anticyclone.

Dew: Condensation of water vapour on surface whose temperature is reduced by rotational cooling to below the dew-point of the air in contact with it.

Dew-point: The dew-point temperature (T_d) of a moist air sample, commonly termed simply the dew-point, is that temperature to which the air must be cooled in order that it shall be saturated with respect to water at its existing pressure and humidity.

Diadelphous: Having two bundles of stamens, *e.g.*, in papilionaceous flowers.

Diakinesis: A substage of prophase I of meiosis; the chiasma terminalization is complete; the bivalents are short and thick, and well-spread in the cell.

Diara land: The lands which are situated in between the natural levees that get inundated for different length of periods braiding and coarse changing of rivers.

Dibbler: A device used for dibbling.

Dibbling: A method of sowing by placing crop seeds in the holes manually by using a dibbler wherein specific spacing and number of plants are maintained between the rows and within the rows.

Dicentric chromosome: A chromosome having two centromeres.

Dichogamy: The maturation of male and female parts at different times.

Dicliny: The condition in which male and female organs are separate, in different flowers.

Dicotyledon (Dicot): Any seed plant having two cotyledons.

Differentiation: The modification of identical cells derived from a zygote into the different cell types; usually associated with organ formation, and characterised by differences in the kinds of genes being transcribed.

Dihybrid: The progeny from a cross between two homozygous parents differing for two genes (or characters); an individuals heterozygous for two genes.

Diluent: Any liquid or solid material used to dilute or carry an active ingredient in the preparation of a formulation or application.

Dimorphic: Having two forms.

Dioecious: Plant species in which unisexual (male and female) flowers occur on different plants; in such species, plants are either male or female.

Diploid: A cell, organism or species having two copies of a single genome; specified by $2x$ (cf. $2n$); or a plant with two sets of chromosomes.

Diploid apogamy: Development of embryos from diploid synergids or antipodal cells.

Diplontic selection: A kind of competition between mutated and non-mutated cells which exists throughout the growth of the mutated plant, and generally, mutated cells are not able to compete resulting in the elimination of the mutated cells.

Diplospory: Type of apomixis in which embryo develops from the megaspore.

Diplotene: The state after pachytene and before diakinesis during prophase I of meiosis; characterised by the movement of homologous chromosomes, paired to form bivalents, away from each other and the appearance of chiasmata.

Direct application: An application of a chemical directed to a restricted area, such as a row or a bed, at base of plants, minimizing contact with the crop.

Direct sprays: It is applied to a particular part of the plant usually to the lower part of the plant stem or truck.

Direct training: Training provided in a situation where the trainers and trainees are in direct face-to-face contact to enable free exchange of ideas and clearance of doubts.

Discontinuous variation: Variation which can be divided into distinct classes; in case of qualitative characters; usually little affected by the environment; yields typical F_2 ratios.

Discordant: When two members of a pair, *e.g.,* twins, show dissimilar phenotypes.

Disease resistance: The ability of plant to withstand, oppose or overcome the attack of pathogen. It is variable in amount ranging from zero to cent per cent and therefore, different degree of resistance, *immunity*: complete resistance; *practical resistance*: High or moderate resistance; *susceptibility*: Slight or zero resistance.

Disjunction: The separation of homologous chromosomes (AI of meiosis) or sister chromatids (A II of meiosis and anaphase of mitosis), and their movement to the opposite poles.

Displaced cropping: When one type of vegetation is overrun by others.

Distant training: Training that happens in a situation where the participants and trainers are widely separated and modern instructional technologies such as radio, television, internet etc. are used to bring interaction among them.

Diuretic: Which cause an increase in the flow of urine.

Divergent question: The divergent questions are inductive and imaginative in nature.

Divergers: Trainees with divergent learning skill rely more on concrete experience and reflective observation.

Diversified cropping: A cropping plan in which no single crop contributes 50 per cent or more towards the total crop production or monetary income (comparable equivalent) annually.

Diversified farm: A farm on which no single product or source of income equals as much as 50 per cent of the total receipt and on such farm the farmer depends on several sources of income.

Diversion drain: A ditch or waterway constructed to intercept and drain the run off from non-arable areas situated in the upper portions of the catchment into the existing drainage systems. These drains help to prevent damage due to overflow, to the arable lands situated in the lower reaches of the terrain.

Diversity: The quantity of different types of organisms, species, cultivators and/or physical elements per unit area.

Diversity index: It measures the multiplicity of farm products which planted by computing the reciprocal of sum of square of the share of gross revenue received from each individual farm enterprise in a single year. Thus, it is in effect an average of the number of crops harvested weighted in particular way by the value of production of each of the crops.

Dizygotic twins: Twins originating from two separate eggs, each fertilized by a separate sperm; may be of the same or different sex (*Syn.*, fraternal twins).

DNA: Deoxyribonucleic acid; the genetic material; serves as the template for DNA replication and transcription; may be unique or repetitive.

DNAase: An enzyme that specifically digests or hydrolyses DNA.

Doka stage: Fruits of date palm when full grown, hard and show yellow or red coloration marks this stage. Datesare harvested at this stage in India.

Dominance: In the heterozygote, one allele expresses itself at the expense of the other.

Dominant: The member of the pair of alleles which suppresses the other member but expressed itself (Intra-allelic interaction).

Donor parent: A parent which donates desirable gene(s).

Dormancy: An inactive period often exhibited by seeds and buds during which growth is stopped.

Dormant spray: A herbicide applied during the dormant period, *i.e.*, after leaf-fall and before bud-break of deciduous plants.

Double cropping: Taking two crops a year in sequence from the same piece of land.

Double cross: A hybrid between two single crosses involving four different inbred lines.

Double crossover: In linkage, the recombinant type produced by simultaneous crossing over on both the sides of a gene; it is the least frequent class in the test cross progeny of a three point cross.

Double dormancy: Seed dormancy owing to both seed coat (lack of water permeability) and dormant embryo.

Double-edged weed killers: A weedicide which will kill monocot as well as dicot weeds.

Double fertilization: Fusion of one of the two male gametes with the secondary nucleus for the development of endosperm.

Double working: The practice of inserting the mutually compatible inter-stock piece between stock and scion to more than two kinds of plants together by grafting.

Downstream: The direction in which the stream flows. It

depends on the direction of the slope.

Dradle knoll: A small knoll formed by earth that is raised and left by an uprooted tree.

Drainage line: A water course into which excess water from the attachment area is discharged. A drainage line combines with other similar lines and flows into the main collection centre, which could be a tank, stream or a pond.

Drizzle: Rainfall in which water drops are very small ranging from 0.2 to 0.5 mm in diameter.

Drop pits: Drop pits are used to check the velocity of run off water. They are dug at a vertical interval of 0.3 metres in drainage lines situated in areas with more than 10 per cent slope. These pits could be made up of stone slabs, old oil drums or wooden planks.

Drosometer: Instrument for measuring dew deposit. It is termed as oetalies dew gauge.

Drought: The moisture deficit which results when the amount of water vapour available in the soil is not sufficient to meet the demand of potential evapo-transpiration.

Drought year: Drought year is that in which rainfall short more than twice the standard of deviation.

Drupe: (*a*) A fruit whose seed is contained in a hard pit or stone covered by soft, pulpy flesh with a thin outer covering. (*b*) A fruit derived entirely from an ovary, one-seeded, with an exocarp, fleshy measocarp, and stony endocarp (as peach, cherry and plum).

Dry air: In physical meteorology this term generally signifies air which is completely dry, *i.e.,* which contains no water vapour.

Dry farming/Dry land farming: The practices of crop production entirely with rain water received during the crop season or conserved soil moisture in low rainfall (less than 800 mm) area of arid and semi-arid climate and crop may face mild to very severe moisture stress during their life cycle.

Dry matter content: Plant substance less water; found by oven drying (usually at 70°C) a weighed sample to a constant weight.

Dry season: A period of a month or more, recurring every year, which is marked in a given region (generally tropical or sub-tropical) by the complete or almost complete absence of precipitation and actual evapo-transpiration falls below potential evapo-transpiration.

Dry spell: A period of at least 15 consecutive days, during which not a single wetting rain has fallen.

Dry water: It is the area irrigated by one-cusec discharge of water during the crop period, it is equal to twice the base divided by delta.

Duplicate: Two or more pairs of genes that produces genes effects whether alone or together.

Duplicate Genes: Two dominant genes produce the same phenotypic effect whether they are alone or together, the alternative phenotype being produced only when both are in the recessive state.

Duplication: The occurrence of a chromosome segment more than once in the same chromosome or in a genome.

Dust storm: An ensemble of particles, dust or sand energetically lifted to great height by a strong and turbulent wind.

Dynamic voice: A voice that represents force and strength, which will attract attention of even sleepy trainees.

E

Earthen spurs: Mechanical measures used to check stream bank erosion particularly at the turning points of the gully. The spurs consist of structures that just inwards to control water flow and direct it along a prescribed course. Spurs are made of earth, stone or masonary depending on the quantum of flow of the runoff.

Earthling up: The process of putting the earth or soil just near the base of stem of certain crops like sugarcane, cassava, papaya, potato, banana etc. to give support to the plants.

Ebracteate: Without bracts.

Eco climate: Climate under crop canopy is called eco-climate.

Eco farming: It is the potential for introducing mutually reinforcing ecological approaches to food production. It aims at the maintenance of soil chemically, biologically and physically the way nature care of plants growing on it. "Feed the soil not the Plant" is the watch and slogan of ecological farming.

Ecological agriculture: Farming practices that enhance or, at least, do no harm the environment and are aimed at minimising the use of chemical inputs, rather than completely avoiding them as in organic farming. Also known as eco-farming.

Ecological crop geography: A branch of crop ecology which deals with the broad spatial distribution of crops/plants and the rationale of such distributions in terms of physical and socio-economic environment influencing the production crops.

Ecology: The study of the mutual relations between organism and their environment.

Economic plants: Plants/trees that serve to bring in subsidiary income to the farmer or act as a buffer if the main crop fails. These plants may yield either fruit, fuel, timber or fodder depending on the farmer's requirement. They are usually grown along keylines and field boundaries and act as good contour demarcators. Examples: Custard apple, Acacia, Dalbergia, Pongamia and Glyricidia.

Economic poison (Pesticides): As defined under the Federal Insectiside, Fungicide and Rodenticide Act, economic poison means "any substance or mixture of substances intended for preventing, destroying, repelling or mitigating insects, rodents, nematodes, fungi or weeds, or any other forms of life declared to be pests and any substance or mixture of substances intended for use as a plant regulator, defoliant or desicent." As so defined, economic poison are now generally as pesticides.

Economic viability: Economic viability of a pattern can be determined by a budget analysis. This analysis includes cost of labour and purchased inputs for all operations specified as well as yields obtained. The profitability and returns to resources (productivity) of the pattern can then be compared with those of the existing cropping system to be replaced.

Economic yield: Economically useful part of total dry matter produced by a plant or crop in unit land area.

Ecosystem: The communities of plants and animals (including humans) living in a given area and their physical and chemical environment (*e.g.,* air, water, soil), including the interactions between them and their environment where there is a continuous flow of energy and matter through the system.

Eco-type: Plant type or strain within a species, resulting from exposure to particular environment.

Edaphic: A term pertaining to the influence or relationship of soil or other similar media to plants in contrast to atmospheric influences.

Edaphic factors: Soil factors.

Edaphology: The scientific study of the relationship between soils and living things, including man's use of the land.

Edible oil: Any vegetable oil which is fit for human consumption.

Education: Education is understood to be the general cultural preparation that a young person receives before entering public employment.

Effective field capacity (tillage): The actual area covered by the implement based on its total time consumed and its width.

Effective irrigation: It is controlled and uniform application of water to crop land in required amount at required time, with minimum cost to produce optimum yields without waste of water and adverse effect on soil in form of soil salinity and water logging problems.

Effective rainfall: It is that fraction of total precipitation which forms a part of the crop consumptive use or that becomes utilisable for crop production.

Efficient cropping zone: It is a cropping zone in which relative yield index is 200 per cent and relative spread indices vary from 90-200 per cent.

Egg (Ovum): The gamete produced by a female organism.

Elasticity of production (EP): Elasticity of production is the percentage increase in output as compared to the percentage in input.

Electrical conductivity (EC): The property of a substance to transfer an electrical charge (reciprocal of resistance) used for the measurement of the salt of an extract from a soil when saturated with water. Measured in milli mhos/cm or dsm^{-1} at 25°C.

Elevation: In meteorology, term used to denote the height of the ground above mean sea level.

Elite variety: An improved variety developed by plant breeders and released to farmers due to it superiority in at least one respect.

Emasculation: The artificial removal of anthers from an immature bisexual flower.

Embankment pond: A body of water harnessed by placing a dam on a well defined water course. Embankment ponds are found mostly in soils with low permeability.

Embryo: The early stages of development of an organism from the zygote. The rudimentary plant within the seed.

Embryo culture: A technique of existing the embryo from the seed at some stage during its development and germinating it on a special medium.

Embryo rescue: A technique employed when offspring of a cross is aborted. In this technique the embryos are excised from the seed and are nurtured of nutrient medium in a sterile test tube until they become a seedling that is strong enough to be planted in soil.

Embryo Sac: In plants, the structure derived from a megaspore through mitosis; generally, contains one egg cell, two synergids, three antipodal cells and two polar nuclei.

Embryogenesis: Morphological and physiological development of the embryo from the single celled zygote.

Emergence: Coming out of the place, such as a germinating seedling breaking through the soil surface or flower from bud or the adult insect leaving the last nymphal skin or pupal case.

Emmer wheat: A tetraploid wheat of the group *Triticum diciccum* which is growing on a limited area in southern India and is preferred in the preparation of suji or rawa.

Empowerment: The act or process of giving power; or state of being empowered.

Emulsified concentrate (pesticide): An emulsified is an oil based liquid compound containing a high concentration of insecticide compound is a mixture of insecticides solvent and emulsifiers to make it easily mixed in water, and wetting and sticking agents to make the material cover and adhere to the plants.

Emulsifier: The material that facilitates suspension of one liquid in another.

Emulsifying agent: A substance which stabilises (reduces the tendency to separate) a suspension of droplets of one liquid in another liquid which otherwise would not mix the first one. Also referred to as an emulsion stabiliser emulsifier.

Emulsion: A dispersion of fine particles of oily material in water (OW type) or less commonly in pesticide usage in dispersion of water droplets in oil (OW type).

Endemic (a plant or animal): Confined to a particular area either because it is evolved there and has not been able to spread more widely or because it had survived there and nowhere else.

Endocarp: The inner layer of the pericarp or fruit wall.

Endogenote: In bacteria, the chromosome segment which is homologous to the exogenote (the DNA segment transferred from a donor cell to the cell in question).

Endomitosis: Two or more chromosome replications not followed by nuclear or cell division; leads to endopolyploidy, *i.e.*, each replication doubles the existing chromosome number of the cell.

Endonuclease: An enzyme which induces internal cuts or cleavages in a DNA molecule.

Endoplasmic Reticulum: A network of membranes in the cytoplasm; consists of tubules, vesicles and cisterns; provides the attachment site for ribosomes; may be rough or smooth.

Endopolyploidy: The condition of cells with twice, four-times, etc. chromosomes of the normal somatic cells of an organism; due to endomitosis.

Endosperm: The nutritive tissue of developing seeds derived from repeated mitotic divisions of the fusion product of the two polar nuclei of an embryo sac with a sperm; usually it is $3n$, but in some cases it is $5n$.

Endospermic seed: The seed in which the embryo is reduced in size in proportion to the rest of the seed, and in these seeds the storage material lies in the endosperm and perisperm.

Endoxerosis of citrus: A physiological disorder of lime and lemons due to faulty nutrition and irrigation where deterioration of the tissues at the styler end of the fruit begins and with its advancement, the acute condition immature fruits are shaded and sometimes cavities are formed inside the tissue.

Energy farming: A concept involving the farming of fast growing plants/trees for the purpose of providing biomass that can be used directly as fuel or converted into other fuel energy products.

Energy productivity ratio: Ratios showing food yield per unit of energy input and are often used in evaluating the energetics of agricultural systems. The energy productivity ratio is food production in nutritional equivalent to fossil-fuel based feedback.

Enthalpy: A thermodynamic quantity which represents the 'total heat' content per unit mass of substance; the units normally employed are joules per kilogram.

Environment: Surroundings, including water, air soil and living organisms and their interrelationships. The sum total of all living and non-living factors outside an organism which affect the development and the functions of the organism.

Enzyme: A protein which catalyses or speeds up a specific chemical reaction; usually in a living system.

Epicotyl grafting: Method of grafting where 7-10 days old seedling stocks are grafted with the current year's scion shoots, which have previously been defoliated, by both splices and cleft methods; practiced in mango, cashewnut, walnut, chestnut etc.

Epigyny: The phenomenon in which the thalamus completely encloses the ovary and getting fused with it, and bears the sepals, petals and stamens on the top of the ovary.

Epinasty: The twisting and curling of leaves and stem caused by uneven growth of cells.

Epipetalous: Borne on or arising from the petals or corolla.

Episome: In bacteria, a DNA molecule, other than the bacterial chromosome, capable of replication; may be integrated into or independent of the bacterial chromosome, it is not essential for the bacterial cells, except under specific environments.

Epistasis: The expression of a gene is modified by another gene; in quantitative genetics, nonallelic interaction or simply interaction.

Equational division: The second division of meiosis when the sister chromatids separate; when crossing over occurs between a gene and its centromere, segregation for the gene takes place during this division.

Equatorial plate: It is the arbitrary plane at which the chromosomes are arranged during metaphase.

Equatorial trough: The shallow trough of low pressure, generally situated on or near the equator, marking the convergence zone of air which moves towards the subtropical anticyclones of either hemisphere.

Equitable system: A fair, just and impartial system.

Equivalent acidity: The numbers of parts by weight of $CaCO_3$ required to neutralize the acidity resulting from the use of 100 parts by weight of fertilizers.

Equivalent conductivity: The conductance of a solution which contains one gram equivalent of the solute when placed between the two electrodes of definite size and 1 cm apart. It is represented by and is numerically equal to specific conductance K multiplied by volume in cc containing one gram equivalent of the solute.

Equivalent weight of soil: The weight of clay or organic soil colloid which has a combining power equivalent to one gram atomic weight of hydrogen.

Eradication (disease): A method of disease control in which the pathogen is eliminated after it has established.

Erodibility: The lack of resistance of a soil to soil erosion.

Erosion: The process of detachment and transportation of soil particles by wind or water. Erosion is influenced by the various factors, *viz.*, Precipitation (The greater the intensity, greater than 25 mm/hr duration and frequency of rainfall the higher the percentage of erosion); Slope (The greater the slope the higher the erosion); Type of soil (The structure, texture, infiltration capacity, permeability and the organic matter present in the soil greatly influence the percentage of erosion); Fine soil is more susceptible than coarse soil. The type of use the soil is put to, also influences the percentage of erosion.

Erosion ravine: An advanced stage of gully erosion. Typical of vast expanses of alluvial soils.

Erosion resistance crop: A crop, because of its dense foliage, root systems etc. provides effective protection against soil loss by erosion.

Erosion rill: In cases where sheet erosion goes unchecked, the silt laden runoff forms well defined and minute finger shaped grooves or channels. This process is termed as rill erosion and can be easily removed by tillage operations.

Erosion splash: Erosion that results from falling rainwater drops. The impact of falling water (rain drops) results in the loosening of soil particles on very wet soils. Subsequently, these soil particles get detached and washed off.

Erosion water: Soil erosion caused by water is the result of application of energy from two distinct sources like falling raindrops, surface flow. Falling raindrops detach the soil particles whereas the surface flow transports the soil. When erosion also strips the fertile topsoil reducing the productive capacity of the soil. Fine soil particles are carried in a series of short bounces called saltations. Very fine particles (0.1 mm diameter) are carried in suspension. Soil particles that are too heavy to be carried are pushed or spread along the surface by the impact of particles in saltations. This movement is termed as surface creep.

Erosion wind: Wind erosion takes place in arid and semi-arid areas devoid of vegetation. The soil particles on the land surface are lifted and blown away by winds with velocity of 4 kmph. When the velocity of the dust-bearing winds is retarded, coarser soil particles are deposited in the form of dunes. Fertile lands are thus rendered unfit for cultivation.

Erosion, gully: Gully erosion is a consequence of unchecked rill erosion. Tiny grooves or rill develop into wider, deeper channels called gullies. The gullies cut large fields into several small fragments and in the course of time make them unfit for cultivation.

Erosion, sheet: Sheet erosion is the removal of a thin cover of soil more or less uniformly from all parts of the slope. Most often the cultivator is unaware of the loss. The loss can be detected by the muddy colour of the runoff. Sheet erosion causes the soil to deteriorate slowly and imperceptibly.

Erosivity: The ability of rainfall to cause soil erosion.

Erythroblastosis Foetalis: The haemolytic anemia of newborn infants; when mother is Rh^- and father is Rh^+, infants may develop this condition.

Espalier: Training of fruit trees by planting them against the wall and main branches are trained in a plane parallel to the wall in a geometric design.

Essential elements: Elements obtained by a plant from the soil and air and without which plant can not complete its life cycle.

Estrogen: The female hormone.

Etaerio: An aggregate of simple fruits borne by single flower having apocarpous pistil (free carples). The ovaries of respective carpel mature into fruitlets clustered together on the pedicel, e.g. etaerio of berries (*Annona squamosa*), of drupes (*Rubus idaeus*) etc.

Etiolation: Development of plant parts or plants in absence of light which results in chlorophyll degradation, more auxin accumulation, and presence of more undifferentiated parenchymous cells in the etiolated sites.

Euchromatin: The parts of a chromosome which take deep stain during cell division but do not stain during interphase; genetically active parts of chromosomes.

Eugenics: The application of principles of genetics for the improvement of mankind.

Eukaryote: An organism whose cells have true nuclei, *i.e.*, nuclei enclosed by typical nuclear envelope.

Euploid: A cell or an organism in which the chromosome number is an exact multiple of the basic number (x) for the concerned species.

Evaluation questions: The technique of evaluation question makes use of divergent question with an additional component of evaluation.

Evaporation: Evaporation is the term used to describe the process wherein moisture escapes in the form of vapour from a free water/moist soil surface by the effect of sunlight. In meteorology, the change of liquid or ice water vapour. The rate of evaporation is controlled by water and energy (mainly solar radiation) supplies and by the ability of the air to take up more water.

Evaporation fog: Fog which is formed by evaporation from the earth's surface and transpiration from the vegetation.

Evaporimeter: Device for measuring the amount of water evaporated into the atmosphere in a given time interval.

Evapo-transpiration: It is the total amount of water lost due to transpiration by crop and evaporation from the soil surface during a specified time from particular areas. It is also referred to as consumptive water use.

Evolution: The development of new varieties, races, species, general etc. through the origin of variation and natural selection.

Exanthema of citrus: A copper deficiency disorder where black or brown raised lesions on the fruit are formed.

Excavated ponds: These ponds are constructed on flat lands or lands with a rolling topography. The ponds may be lined with plastic, cement or brick. Excavated ponds can provide crop saving irrigation to limited area. They also function as a source of drinking water for human beings and livestock.

Exchangeable cation percentage (ECP): This term indicates the degree of saturation of the soils exchangeable with a cation.

Exchangeable phosphate: The phosphate anion reversibility attached on the surface of the solid phase (anion-exchange capacity) of solid, in such from that it may go into molecular solution by anionic equilibrium reactions with other anions of the solution phase.

Exchangeable potassium: The potassium which is held mainly by the colloidal portion of the soil and is easily exchanged with cation of the neutral salt solutions. It is readily available to growing plant roots.

Exchangeable sodium percentage (ESP): It is the degree of saturation of the soil exchange complex and expressed in percentage.

Exocarp: The outer layer of fruits derived from an ovary as with the skin of a peach.

Exogenote: In bacteria, the chromosomal segment transferred from a donor cell to a recipient cell and homologous to a chromosomal segment of the latter (*cf.* endogenote).

Exon: The expressed sequence of an eukaryotic gene; the part of an eukaryotic gene which is represented in the mature mRNA and in the polypeptide specified by the gene.

Exonuclease: An enzyme which digests DNA or RNA molecules beginning at their ends.

Exotic (plant): A newly introduced plant not native to a place.

Expected Ratio: In chi-square test, the ratio in which the data are expected to occur in a given experiment; based on known facts and theoretical considerations.

Experiencing: This is the first stage of experiential learning cycle which deals with generation of data as a result of going through activities which is a part of structured experience.

Experimental design: A logical structure of an experiment, that helps in obtaining results with some precision.

Experimental yield: The yield of a crop/variety obtained at experimental stations where yield maximization is the major objective using all sophisticated technology.

Experimentation: The process in microlab in which the participants are provided with opportunities for experimentation and experiences.

Expressive voice: This voice that beats away monotony and arouses the emotion of trainees.

Expressivity: The degree of manifestation of a genetic character or the degree of phenotypic expression of a gene in the different individuals; it may be uniform or variable.

Ex-situ: Outside the natural habit.

Extension: Refers here to agricultural extension activities that disseminate research findings and advice to farmers on agricultural practices and improve farmers analytical capacity and communication so as to help them in their decision making related to farming.

Extension agricultural: Agricultural extension is a service which assists farmers through educational procedures related to improving framing methods and techniques, increasing production efficiency and improving their general standard of life.

External inputs: Inputs that originate from outside the system (farm, village, region, country). Artificial external inputs are inputs that need high quantities fossils to be produced or distributed, *e.g.*, synthetic fertilisers, pesticides, pumped irrigation water.

External training: Advanced training provided to an employee by way of deputing to other training institutions.

Extinction coefficient: The ratio between the light losses through the leaf to the light at the top of the leaf.

F

F Factor: In bacteria; fertility factor which converts a cell into male; cells with *F* factors have characteristic *F*-pili.

F_1: The progeny obtained by crossing or mating two strains or individuals; first filial generation.

F_2: The second filial generation; obtained by self-fertilization of or intermating among the individuals of F_1 generation.

Factorial design: An experimental design in which at least two series or treatments are used being combined in every possible combination.

Fallow: It is the practice of allowing cropland to lie idle during growing seasons to build up the soil moisture and fertility content so that a better crop can be produced the following year. It is usually worked periodically to control and improve moisture infiltration; it is often colonised by natural vegetation and may be grazed.

Fallow land: Land left unframed for one or more growing seasons to kill weeds make the soil richer.

Fallow system: It is the sequence of crop years with fallow years. Extensive fallow systems are shifting cultivation systems. Intensive fallow systems are bush fallow and grass fallow systems.

Farm: An area of land used for agriculture either to raise a crop or pasturage or maintain livestock.

Farm forestry: Farm forestry is a process in which trees are grown specifically for fuel, food and for variety of other resources on farm. It is a "tree dimensional farming" and provides an alternative to monoculture grain farming. The three dimensions are defined as the trees themselves, the harvest from the trees which is used to feed livestock and the animals and the animal products.

Farm planning: A process involving many decisions to be taken in respect of the kinds of crops to grow, rotations, mixtures, soil and water conservation practices to be followed and building, bullocks, machinery purchase etc.

Farm pond: Small bodies of water made either by constructing an embankment across or a drainage course or by excavating a pit or a dugout. In arid areas where the water table is usually very deep, surface runoff is the only source for feeding these ponds which are made 'pucca'. Locally they are known as 'Tankas'. Farm ponds serve as sources of drinking water, or as ideal locations for aquaculture and/or for supportive irrigation for a few fruit trees.

Farm system: All the components with a given farm boundary which interact as a system, including people, crops, livestock, other vegetation, wildlife, and the social, economic and ecological interactions between them and with the environment.

Farm yard manure: Comprises mostly of animal excreta and cattle shed wastes. It is applied in a partially decomposed state to the field before sowing. Farm yard manure contains, most of the essential nutrients to a fair degree. The average percentage distribution of N, P and K of farm yard manure is about 0.5 to 1.5 per cent of N, 0.4 to 0.8 per cent of P_2O_5 and 0.5 to 1.9 per cent of K_2O.

Farming system: Farming system is an integrated approach to farming with several enterprises (*e.g.*, rice cultivation, pisciculture, horticulture, poultry and cattle dairying), designed to be complementary to each other. A combination of one or more enterprises with cropping, when carefully chosen, planned and executed, gives greater dividends than a single enterprise, especially for small and marginal farmers.

Farming system research (FSR): An applied approach to agricultural research conducted by multidisciplinary teams who assess the scope for, and potential impact of, technology changes with a holistic farming system framework. The major steps consist and of identifying homogenous groups of farmers within specific agroclimatic zones; identifying problems and opportunities of these clients; designing new technologies to suit their conditions; testing these innovations in on farm trails, evaluating the results and recommending successful innovations for wider dissemination.

Fasciation: The condition of several stems fused together in a single plane, causing abnormal and deformed growth.

Fasicle: A condensed or close cluster, as of flowers.

Feedback inhibition: Inhibition of an enzyme of a biosynthetic pathway by the end-product of the pathway when it accumulates in the cell upto the threshold concentration; usually the first enzyme of the pathway is involved.

Feni: A fermented product of the juice from cashew apple.

Fertigation: Application of fertilizers with irrigation water. Its applicability depends besides the method of irrigation used, on the type of soil and crop, the climatic conditions or water quality and the types of fertilizers available.

Fertilisation: The fusion of the male and female gametes.

Fertilisation (soil): The application of fertilisers to the soil that aid in the nutrition of plants. A material in which declared nutrients are in the form of inorganic salts obtained by extraction and/or by physical and/or chemical industrial process (also termed as mineral fertilizers).

Fertiliser requirement: The quantity of certain plant nutrient elements needed in addition to the amount supplied by the soil, to increase plant growth or crop yield to a designated optimum.

Fertilisers: Fertilisers are any organic or inorganic material of natural or synthetic origin that are added to soil to supply nutrients essential for plant growth. Fertilisers can be broadly classified into organic, inorganic and bio-fertilisers.

Fertilisers, bio organic: Biofertilisers are bacteria and other microorganisms, that are capable of fixing atmospheric Nitrogen and are also used to inoculate seeds, particularly for pulse crops. The use of biofertilisers by growing legumes with non-legumes in the cropping system is an age-old agronomic practice. Organic fertilisers contain carbon in large quantities and are derived from plant and animal residues.

Fertilisers, inorganic: These are also known as chemical or synthetic fertilisers. They are of concentrated nature with one or more nutrients in readily available forms. Some of the most common fertilisers and their ingredients are given below:

Fertilisers	*Percentage of ingredients*		
	N	*P*	*K*
Urea	46	0	0
Single Superphosphate	0	16	0
Muriate of Potash	0	60	0
Diammonium phosphate	11	46	0
Mixed Fertilisers	15	15	15

Fertility: The ability of an organism to produce progeny.

Fertility (soil): The ability of a soil to supply all the essential nutrients in optimum amount and balance and in a form readily available to plants.

Fertility gradient: The variation in natural fertility in any direction across an area of land.

Fibrous Root System: The root system that is composed of profusely branched roots with many lateral rootlets but with no main or tap root development.

Field capacity: It is the moisture content in percentage of a soil on oven dry basis when it has been completely saturated and downward movements of excess water has practically ceased, which usually takes place within 2-3 days after saturation. It is the upper limit of available soil moisture for plant growth.

Field crops: Herbaceous plants grown on cultivated fields.

Field dependence learners: Learners who perceive information in holistic fashion.

Field efficiency: It is the ratio of effective field capacity and theoretical field capacity expressed in percentage.

Field heat: It is the heat of the produce after harvesting from fields which can be determined by the thermal energy removed from the produce in cooling.

Field independence learners: Learners who analyze the study materials into different parts and like to work independently.

Field irrigation efficiency: It is the ratio expressed in the percentage of irrigation water available for use of crops to that delivered to a field.

Field irrigation requirement of a crop: The per cent moisture content in a field sample of soil at any one time.

Field strip cropping: A system of strip cropping in which crops are grown in parallel strips laid across the general slope but which do not follow the contour. Strips of grass or close-growing crops are alternate with field crops, often.

Field systems: It is practiced where one arable crop follows another and where established fields are clearly separated from each other.

Field water capacity: The maximum amount of capillary water that a particular soils is able to hold.

Fifteen-atmosphere percentage: It is the moisture percentage on dry weight basis of a soil sample which has been wetted and brought to an equilibrium in a pressure membrane apparatus at 15 atmospheric pressure (221 1b/sq. inch). This characteristic moisture value for soils approximate the lower limit of water available for crop growth (permanent wilting percentage).

Fifty-percentage yield decrement value: The measured value of the soil salinity/alkali that decreases crop yield by 50 per cent as compared with yields of the same crop on non-saline and non alkali soils under similar growing conditions.

Filament: The stalk of a stamen.

Filial (F_1, F_2, etc.): Related by descent, F_1 – the first Filial generation of a cross, F_2 – the second Filial generation and so on.

Filler: Any materials mixed with fertilizers for any purpose other than the addition of available nutrients, such as for conditioning to give anti-caking properties and for increasing the weight to bring the percentage of nutrients to desired values.

Fitness: The number of progeny produced by an individual or a genotype in relation to the average of the population or to that of the best fit genotype.

Fixation: The frequency of an allele in a population attaining the value of one.

Fixation (nutrient): It is a process in a soil by which certain chemical elements essential for plant growth are converted from a soluble or exchangeable form, for example, phosphate fixation.

Fixation of nitrogen: The fixation of atmosphere nitrogen in the soil by symbiotic and non-symbiotic bacteria.

Fixation of phosphorus: The conservation of soluble phosphorus nutrient in the soil into less soluble and unavailable forms.

Fixed phosphorus: (*i*) That phosphorus which has changed to less soluble forms as a result of reaction with the soil; moderately available phosphorus. (*ii*) Applied phosphorus; phosphorus not taken up by plants during the first cropping year. (*iii*) Soluble phosphorus, phosphorus attached to the solid phase of soil in forms which is highly unavailable to crops. (*iv*) Unavailable phosphorus in other than readily or moderately available forms.

Fixed potassium: Potassium held by a soil which is neither water soluble nor readily exchangeable.

Flavedo: Outer yellow coloured portion of citrus peel in which oil glands are embedded.

Floating blanket: A donoe crust of algae floating in the water ponds.

Floating rice: Tall rice grown where maximum water depth ranges between 1-6 m for more than half of the growth duration.

Flute budding: A method of budding in which a patch of bark is removed leaving a narrow connection between the upper and lower parts of the stock; a bud patch of same size from the bud stick is removed and fitted to the stock and wrapped with polythene sheets.

Fodder crops: The cultivated plant species that are utilized as livestock feed in the form of soolage, silage and hay.

Fog: Obscurity in the surface layer of the atmosphere, which is caused by a suspension of water droplets, with or without smokeless particles, and which is defined by international agreement, as being associated with visibility less than 1 km.

Fog precipitation: Fog precipitation, sometimes also termed 'fog drip' signifies the precipitation of liquid water from non-bearing could due to the interception of the cloud particles by trees and other vegetation.

Follow-up: A follow-up of a training programme can be defined as activities related to finding out the utility of the learning outcome of the programme in job performance, analysis of reasons for non-utilization of new skills or knowledge acquired, providing help and reinforcement for application of learning as well as assessing future training requirements.

Foot-candle: The light intensity of one standard candle at a distance of one foot from a light source.

Forage crop: Crops grown primarily for livestock feed, to be either harvested for hay, silage or green feed or harvested by grazing animals.

Forage forestry: The concept of forage forestry can simply be defined as "using the same land or portion of it, simultaneously or sequentially for fodder, food and fuel".

Forecast: The term, first applied in meteorology by Admiral Fitz Roy, which signified a statement of anticipated (meteorological) conditions for a specified place (or area, route, etc.) and period of time.

Forest: A plant community predominantly of trees and other wood vegetation usually with a closed canopy.

Forestry: The act, occupation or art of farming and cultivating forests and systematic utilization, reproduction and improvement of the productivity capacity of trees including the planting and culture of new forests. Efficient forestry programmes can help to improve soil quality and are recommended from class V land onwards. Of the 100 million hectares of non-arable land in our country, nearly 70 million hectares are in some state of degradation. Desertification is creeping in forestry. Measures must, therefore, be undertaken urgently to stabilise the ecological balance, and improve the status of the degraded land.

Forestry, afforestation: Afforestation is the growing of forests on non-arable land (class V to VII) which has been previously unproductive. Forestry, Avenue plantation: This refers to the planting of trees along the roadsides. They provide shade, green the area and eventually improve the fertility of the soil. These trees also act as windbreaks. The most common species include Neem, Tamarind, Eucalyptus, Casuarina and Mango.

Forestry, block plantation: This refers to the planting of trees, either forest species or horticulture plants or both over a large area. Block plantation is usually taken up for commercial purposes. Areas under class V and above, which have severe limitations for crop production are ideal for block planting. The species grown commonly are Eucalyptus, Casuarina, Acaci, Bamboo, Mango, Neem, Teak and Tamarind.

Forestry, off-shore plantation: It is the planting of trees in areas situated between tanks and cultivated fields. This area gets a lot of the accumulated soil from the catchment and is very fertile. Species that can withstand waterlogged conditions should be chosen for off-shore plantations. Economic plants like Acacia which are useful for fuel can also be grown.

Forestry, reforestation: It refers to tree planting activity taken up in denuded forest land of class V to VII where vegetation is sparse, by once again raising forests.

Forestry, social: It refers to the programmes of forestry development and conservation programmes under various agro-ecological conditions. It includes mixed plantations on wastelands and Panchayat lands. Reforestation of degraded forestland, and raising of shelter belts among others. The objective of social forestry programmes in rainfed areas is to green the area, to meet the food, fodder and fuel requirements of the community, to provide gainful employment to the farmers or landless labourers in the area, and to protect cultivated fields from water and wind erosion. Social forestry schemes can be taken up on government or community land. Most social forestry programmes are afforestation programmes, as they involve planting of trees where there were none.

Forkert budding: Method of budding where a transverse cut and two vertical cuts joining the transverse cut on the stock are given and the bark is carefully peeled along these cuts but kept attached on the lower side. The desirable bud patch is then fitted into the exposed portion of the stock and covered with the flap of the bark and wrapped with polythene tape; practiced in peach, pear, etc.

Formal training: Training programme conducted at a particular place with a specific syllabus and schedule is considered as formal in modality.

Formulation: A mixture of an active pesticide (herbicide) chemical with carriers, diluents or other materials; usually to facilitate handling.

Fossil: Any impression, natural or impregnated, ramming, or other traces of an animal or plant of past geological ages which has been preserved in the earth's crust.

Foundation block: A planting of fruit trees in which stock is maintained permanently under rigid control to be used as a primary source for "true-to-type" stock and not as a direct source of propagating material itself.

Foundation seed: The second look in the certified seed chain produced from breeder seed and handled in such a way as to ensure genetic identity and varietal purity.

Foundation training: The purpose of foundation training is to strengthen the basic foundational knowledge needed by an employee which includes rules and regulations, financial transactions, administrative procedures, communication and leadership skills apart from core subject matter.

Founder Principle: The theory that small isolated populations diverge, in terms of gene and genotype frequencies, from the parent population due to sampling error at the time of the establishment of these populations.

Frameworking: A method of changing the cultivar of fruit trees where all the main scaffold branches are retained and most of the small laterals are replaced by a large number of scion of the replacement cultivar.

Free board: It is the vertical distance between the highest water level anticipated in the design of an irrigation structure (reservoir dam, canal, channel etc.) and the top of the retaining bank.

Freezing injury of citrus: Typically a collapse and subsequent desiccation of affected internal tissues with little or no external symptoms in the fruit excepting rind thickening.

Frequency distribution: Refers to data classified on the basis of some variables, which can be measured on a scale.

Fruit: The ripened ovary of group ovaries together with other adhering structures.

Fruit drop: Dropping of fruit at different stages fruit development due to competition among the growing fruitlets, lack of pollination and fertilization, unfavourable environmental conditions, hormonal misbalance etc.

Fruit pit: A disorder, more especially of apples, in which pits of fruit tissues become darkened in colour, dry out and usually bitter in taste.

Fruitfulness: The capacity of flowering and producing fruits in abundance.

Fumigant: A volatile liquid or gas to kill insects, nematodes, fungi, bacteria, seeds, roots, rhizomes or entire plants; usually applied to soil or in an enclosure of some kind.

Functional diversity: The quantity of different organisms, species or cultivars that contribute to increasing the stability, productivity or continuity of an agro-ecosystem.

Fungi: Plural of fungus; any of a large group of plants (including moulds, mildews, mushroom's, rusts and amuts) which are parasites on living organisms or feed on dead organic material, lack chlorophyll, roots. Stem and leaves, and reproduce by means of spores.

Furrow: A trench formed by the plough in the soil during tillage operations.

Furrow, dead: In low rainfall areas where contour farming is practised deep furrows are formed along the contour at 2-3 metre intervals.

Fusiform: Tapering towards both ends.

G

Game: A game is a structured activity in which two or more participants compete within constraints of rule to achieve an objective.

Gamete: A specialized haploid (usually) cell which participates in sexual reproduction; one male and one female gametes fuse to produce a zygote.

Gametic chromosome number: The chromosome number normally present in the gametes of a species; ordinarily, it is the haploid number (n).

Gametogenesis: The production of male and female gametes.

Gamopetalous: Petals united together at least at the base.

Gamut-running game: Games, which involves overcoming obstacles by the players until they reach a goal.

Garden agriculture: Crops grown on land more or less adjacent to settlements; usually vegetables etc. rather than staples dependent on organic wastes from settlement for nutrient inputs.

Gender: The socially constructed roles and responsibilities assigned to women and men in a given culture to location and the societal structure that supports them.

Gender analysis: A close examination of a problem or situation to identify gender issues.

Gene: A segment of DNA which specifies a single polypeptide (*structural gene*); also, a segment of DNA which codes for an RNA molecule (*e.g.*, rRNA and tRNA genes), or may serve as a binding site for regulatory proteins (*e.g.*, operator and promoter genes).

Gene conversion: A process of non-reciprocal recombination due to which an allele is present in an excess of the other in the recombination products.

Gene frequency: The proportion of one allele relative to all the alleles of a gene in a Mendelian population.

Gene pool: The sum total of all the alleles present in the breeding or reproductive members of a random mating population.

Gene sanctuary: An area within the centre of diversity of a particular plant species which is protected from human interference.

Gene symbol: A letter or a group of letters used to denote a gene; usually; derived from the name of the mutant phenotype; the first letter of the symbol for the dominant allele is capital, which all the letters in the symbols for recessive alleles are in lower case.

General system games: Games related to total system of an organization, i.e. an institution or a company under a highly structured set-up.

Generalizing: The final stage of experiential learning cycle where the participants are helped to relate a classroom or training room experience to the real world experience.

Genetic block: The block in a biochemical reaction due to a mutant allele of the concerned gene; the concerned enzyme produced by the mutant allele is defective or non-functional.

Genetic drift: *See, Random Genetic Drift.*

Genetic erosion: Disappearance of genetic resources.

Genetic fine structure: Essentially a linkage map of the various mutant alleles of a gene; based on complementation and recombination data.

Genetic load: The proportion by which the fitness of a random mating population decreases as compared to that of the optimum genotype due to the presence of deleterious alleles in the population.

Genetic resources: Plant and animal stock with different inheritable characteristics of (potential) use within an agro-ecosystem.

Genetics: Science dealing with similarities and variations among related individuals. The branch of biology concerned with the study of heredity and variation.

Genome: A complete set of chromosomes of a diploid species;

all the members of a genome are distinct from each other in gene content and, often, in morphology; members of a genome do not pair.

Genotype: The genetic constitution of an organism.

Genus: A group of related species or sometimes a single species, Plural-genera.

Geopotential: The potential energy per unit mass of body due to the earth's gravitational field referred to an arbitrary zero.

Germ Cell: A cell destined to produce gametes for sexual reproduction.

Germination: A process of growing, developing or sprouting. It is the resumption of growth of a seed usually re-organised by ruptures of the seed coat or spore wall and appearance of the radicle and plumule from the seed.

Germplasm: The hereditary material transmitted from parents to offspring through germ cells.

Girdling: Process of removing a strip of bark from branches or stem which blocks the downward translocation of carbohydrates, hormones beyond that constriction and rather accumulate above it.

Glabrous: Non-hairy, smooth, hairless.

Glaucous: Covered with a whitish substance, called bloom.

Glycoside: One or a group of compounds yielding sugar and another substance on hydrolysis.

Gonad: The organ concerned with the production of gametes, *e.g.,* ovary and testis; in case of animals.

Gossypium arborium: A diploid old world cotton species within 13 chromosomes which is widely grown in rainfed areas. It includes perennial or annual shrubs, 2-10 ft. high. The centre of origin is believed to be Bengal region of Asia. Ginning percentage range from 24 to 38 per cent.

Gossypium barbadense (Sea island cotton or Egyptian cotton): A tetraploid new world cotton species within 26 chromosomes. The centre of origin in tropical South America. Ginning percentage range from 32 to 34 per cent.

Gossypium herbaceum: A diploid old world cotton species within 13 chromosomes, constitute a fairly large percentage of medium staple cotton grown in India. Compared to arborium cotton, these are of longer duration and better suited but the upper limits of yield lint length, fibre weight and fibre maturity in this species are inferior to arborium types. Ginning percentage varies from 27 to 36.5 per cent.

Gossypium hirsutum (American cotton or upland cotton): A tetraploid new world species within 26 chromosomes. The centre of origin is known to be Central America. Ginning percentage varies from 33 to 37 per cent.

Graded terrace: A terrace having a constant or variable grade along its length.

Graft chimera graft hybrid: Plant composed of tissues in intimate association from two different individuals which originate by grafting.

Graft incompatibility: Inability of stock and scion to form a successful graft union.

Grafting: (*a*) The process of putting a shoot or bud from one tree or plant into slit in another tree or plant, so that it will grow as a part of the host. (*b*) A from of asexual propagation in which a scion or bud plant is placed in innate contact with another until they grow together.

Grain: An indehiscent simple fruit of a grass in which the seed coat is united with the pericarp.

Grain legumes: These crops belonging to the family leguminosae which are grown for their edible seeds.

Granulation: A disorder in citrus wherein the juice vesicles becomes hard and greenish which causes decrease in soluble sugars, acids, and accumulation of polysaccharides and increase in sodium, calcium, and magnesium content. It is influenced by variety, rootstock, climate, nutrition etc.

Grass: Botanically, any plant of the family *Graminaceae* is called grass. The term 'grass' in the context of feeding livestock and grass land agriculture is not limited to the narrow botanical sense alone, but also includes their common associates of the legume family. However, the term does not include cereals when grown for grain.

Grassland: It is an ecological community in which graminaceous (grasses) species predominate the vegetation. Trees and shrubs are either altogether absent or present in relatively small numbers. According to the landuse statistics, the area under grasslands and permanent pastures is 13 million hectares. Grasslands, the cheapest source of animal feed are in a degraded or denuded state because of faulty management over the ages. Added to this is the fact that they are at their best only during the monsoons after which their condition deteriorates.

Grassland ecosystem: It pertains to the ecosystem of grasslands and pastures.

Grassland farming: Farming system that emphasis the importance of grasses and legumes in livestock and land management in which the legumes are the keystones and grasses from the backbone of the grassland farming.

Gravitation water: It is the excess water above field capacity in a soil moving down under the influence of gravity.

Grazing: The eating or partial defoliation of any kind of standing vegetation by domestic livestock or other animals.

Grazing intensity: The number of animals per unit area of grazing land.

Grazing management: The proper management of grazing animals over a grassland so that the vigour and productivity of the pastures is maintained. Care should be taken to ensure that—there is enough grass left in a grazing ground or a pasture after the animals have grazed. This ensures the soil protected, there is proper usage of forage. This can be done by regulating the grazing intensity.

Grazing, continuous: This is the most widely practised method. Here, the animals graze the same area throughout the year. The main disadvantage of this method is the continued overuse of preferred plants. As a result palatable and nutritious fodder species may give way to coarser grass varieties causing deterioration of the pastures. The method can be adopted with good results if the number of animals are limited to the optimum and if the animals are withdrawn from the pasture before the end of the growing season.

Grazing, deferred rotational: This is slightly different from rotational grazing. In this case a portion of the pasture is left ungrazed, until seed maturity. A different portion is left untouched, each year, over the rotation period of about 3-4 years. The grasses in the rested pasture gain vigour, produce seed and store food in their roots. The rested pasture is then either opened for grazing, or the grass is harvested and preserved for the lean months.

Grazing, rotational: This is a system where the pasture is divided into a number of compartments. Grazing is done in the first compartment for a few days and then the animals are moved to the next compartment. Once the animals have fed on all the compartments they are moved to the first compartment again. Thus each compartment is provided with a growth period in which it can rejuvenate. Rotational grazing improves the productivity of the compartments. It however requires additional investment for the setting up of the compartments. The system has several advantages. It ensures that the pastures have enough re-growth period. It also ensures uniform grazing over the entire area and in the long run produces more fodder than the continuously grazed pastures.

Greenhouse: A house covered with glass used for the protection and cultivation of plants.

Greenhouse effect: The effect of short wave radiation can easily pass through the atmosphere to reach the earth's surface while the outgoing radiation from the earth are not permitted to pass through the atmosphere especially it is cloudy.

Green leaf: This refers to green leaves and tender green twigs collected from shrubs and trees grown on bunds, wastelands and nearby forest areas. The common shrubs and trees useful for this purpose are glyricidia (*Glyricidia maculata*) Sesbania species, karanj (*Pongamia pinnata*), Subabul (*Leucaena Leucocephala*) etc.

Green leaf manuring: Perennial shrubs are specifically planted for this purpose along roads and wastelands, and their leaves are brought onto the field to be ploughed into the soil. It is the similar to green manuring. One tonne of leaf, on an average, yields about 5 kg of Nitrogen. The common species of plants grown for green manure are, *Gliricidia maculata, Pongamia vitex* etc.

Green manure: Any crop plants grown and ploughed under when succulent to improve the soil especially by addition of organic matter.

Green manure crop: A crop grown and ploughed down while still green to add dry matter and nitrogen to the soil.

Green manuring: Green manure is plant material grown on the field and incorporated into the soil when the crop is about to flower green or soon after maturity. The most common species of green manure that are grown are Sunhemp (*Crotalaria juncea*). Dhaincha (*Sesbania aculeata*), Sesbania (*Sesbania speciosa*), and Wild indigo (*Tephrosia purpurea*).

Green revolution: The use of package of inputs, including modern varieties, pesticides, fertilisers and irrigation, in an attempt to increase farm yields in developing countries.

Gross productivity: Total productivity of a system or plant per unit area and per unit time.

Ground truth verification: It is a process of verification of the actual condition of the land with what has been depicted on the topographical map, to note if there have been changes in the general land characteristics.

Ground water: Subsurface water (in the zone of saturation) that is free to move under the influence of gravity is called ground water. It is retained at all levels below the water table. At this level the soil and the underlying strata are saturated with water.

Group discussion: A group discussion refers to exchange of ideas and thoughts by two or more people on a selected topic with a purpose of clarification of ideas or help in understanding and application of ideas to practical situation.

Group interaction games: Games related to human relation problem, which help the participant to experience new points of view.

Growing degree day: Sum over the growing season of a crop of the difference between the daily temperature and a reference temperature.

Growing season: Period of the year during which the growth of vegetation proceeds.

Growth: An irreversible process in which there is an increase in size or dry weight or volume of an organism.

Growth analysis: Method of analysing growth through plant sampling technique. By this method preservations are recorded periodically on various plant characteristics (height, flowering time, ear hearing, tillers, etc.) and also sampled at varying intervals (weekly or fortnightly) when the different parts such as stem, root leaf, fruit seed etc. are separated, weighed and measured. From these the physiological parameters such as growth rate are calculated and these are further analysed in term of contributing component parameters like net assimilation rate and leaf area.

Growth correlation: The different parts of a plant do not grow in isolation but fallow an orderly pattern due to control of one part of the plant over the other. This controlling effect exerted by one part over the other is known as correlative effect.

Growth curve: A graph in which the size of some plant characteristics (height, weight, leaf area etc.) is plotted against time or age. The rates of growth of any single cell or mass of cells are sigmoid (shaped), slow at first, increase with time to maximum and then fall off at about the same rate to zero. The period of maximum growth is called "grand period" of growth.

Growth factor: A specific substance that must be present in the growth medium to permit cell to multiply.

Growth medium: Any materials such as soil, peat etc. used as a support for plant roots, that has a capacity for water retention and which may contain added and or naturally occurring nutrients.

Growth regulators of growth substances: It is a compound, other than nutrients, which in small amount promote, inhibit or otherwise modify physiological process in plants.

Growth stages of cereal crops: (*i*) Tillering stage—Additional shoots are developing from the crown; (*ii*) Joining stage—When stem interposed are elongating; (*iii*) Boot stage—When leaf sheath swells due to the growth of developing spike; (*iv*) Heading stage—When seed head is emerging from the sheath.

Growth substance: A growth regulating chemical produced by a plant of a synthetic one with activity even at very low concentration.

Guard cropping: Means growing main crop in the centre surrounded by hardy and thorny such as safflower around pea, mesta around sugarcane and sorghum around maize.

Gully: The channel formed due to an advanced stage of erosion caused by factors like a high percentage of runoff and moving icebergs. A gully is characterised by the severe loss of top soil and is unfit for cultivation.

Gully erosion: Removal of soil by running water, with formation of channels that cannot be smoothed out completely by normal cultivation.

Gully plug: A gully plug is a mechanical or live structure, placed across the gully, to check the flow of runoff. The purpose of the gully plug is to: increase percolation, reduce the velocity of runoff flow, reduce further cutting of the soil bed, and stabilise and regenerate the gully bed in course of time.

Gummosis: A general disorder of stone fruits, in which exudation and deposit of gum occurs.

Gust: A rapid increase in the strength of the wind relative mean strength obtaining at the time. It is much short-lived than a squall and is also different in nature.

Guttation: The exudation of water from the ends of the vascular system at the margins of leaves under humid conditions.

Gynandromorph: An individual in which a part of the body is male, while the rest of it is female; usually in *Drosophila*; mosaics for the sex chromosome; male part (*XO*) may range from a small section of the body to half of the body.

Gynoecium: The female parts of a flower.

Gynomonoecious: A sex form where pistillate and hermaphrodite flowers are separately produced in the same plant.

Gynophore: The portion below the ovary which grows into a peg like structure as a result of the stimulus of fertilisation.

H

Habitat: The environment in which a plant or animal's lives and which responds to its specific needs.

Hail: Solid precipitation in the form of balls or pieces of ice (hailstones) with diameter ranging from 5 mm to 50 mm or even more.

Hailstorm: A storm, often prolonged and severe, consisting largely of hail of frozen raindrops ranging in diameter from 5 mm to 50 mm or more, the ice particles are precipitation either separately or as aggregates of irregular size and shape.

Hairy hygrometer: Hygrometer in which sensing element is a strand or strands of hair, the change of its length is a function of humidity of the air.

Halophytes: Plants tolerant to high soil salinity conditions.

Haploid: A cell or an organism having the gametic chromosome number.

Harbage: Leaves, stem and other succulent parts of forage plants upon which animal feed.

Hard water: Water which contains minerals, usually calcium and magnesium sulphate, chlorides or carbonates, in solution, to the extent of causing a curd or precipitate, rather than a later, when soap is added; hard water may cause precipitation in some herbicidal sprays.

Hardy-Weinberg Equilibrium: In a random mating population, a gene with two alleles exhibits p^2 *AA* : $2pq$ *Aa* : q^2 aa genotypic frequencies (where, $p + q = 1$).

Hardy-Weinberg Law: In random mating populations, gene and genotype frequencies remain constant after generation provided there is no migration, mutation, selection or random drift.

Harrowing: A multipurpose operation for crushing clods, levelling the seed bed and destroying the germinating weeds while bringing the soil to a fine tilth. The implement used is the blade harrow. Harrowing breaks the clods formed after deep ploughing. It aerates the soil enabling it to absorb and retain more rainwater.

Harvest index (H.I.): Yield of the plant parts of economic interest (economic yield) as percentage of total biological yield in terms of dry matters.

Harvesting: The operation of cutting, picking, plucking or digging, or a combination of these for removing the useful part or economic part from the plants.

Haustorium: The major organ of parasitic weeds for attachment and penetration of the host tissue.

Hay: Hay is feed produced by dehydrating green forage to a moisture content of 15 per cent or less that the biological processes do not proceed rapidly enough to build up heat.

Haylage: Product resulting from ensiling forage with 35-50 per cent moisture under aerobic conditions, compared to hay, it greatly reduces the time involved in field curing and the associated field risk.

Heading back: Removal of a portion of a stem leaving another portion for promoting new growth.

Heat unit: Accumulated degree hours above a certain base temperature specified for a particular crop which serves as a guide to predict the harvest.

Heat wave: The heat wave is defined as appreciable rise in maximum temperature and criteria for describing heat wave are as follows:

Heat wave	3 to 4°C
Above normal	2°C
Appreciable above normal	3 to 4°C
Moderate heat wave	5 to 6°C
Nomenclature	*Departure from normal*
Normal	–1 to 1°C

Severe heat wave 5°C or more
Severe heat wave 7°C and above

(*i*) When maximum temperature remains 45°C or more for two days or more the condition may be described as heat wave.

(*ii*) When normal maximum temperature of the area/station is 40°C or less.

(*iii*) When normal maximum temperature is more than 40°C.

Hela: High-external-input agricultural, depends on significant levels of natural or, more commonly, artificial inputs, such as fertilisers, pesticides and fossils energy, which originate from outside the system (farm, village, region, country) and generally have to purchased.

Heliophytes: Plants of sun loving species, require intense light for normal development.

Heliotrophic: Turning towards the sun.

Hemizygous: A diploid organism having only one copy of a gene in place of the normal two, *e.g.*, in case of sex linkage and in deletion heterozygotes.

Hemophilia: The tendency of bleeding freely from even minor cuts due to an inability of the blood to clot; a genetic disease due to a sex-linked recessive gene.

Hen and chicken disorder: A physiological disorder of grapes where only a few normal sized fruits are matured in a bunch and the rest being undersized and worthless, caused due to the deficiency of boron.

Herbaceous: Like a herb; having stems that are usually soft and not woody.

Herbaceous plant: A vascular plant that does not develop woody tissue.

Herbicide: (*i*) A chemical used for killing or inhibiting the growth of unwanted plants. (*ii*) A class of pesticides that destroys or reduces the negative effects of weeds.

Heredity: The transmission of traits (more precisely, genes) from parents to offspring.

Heritability: The extent of transmissibility of a character. The ratio of genotypic variance (*brood sense heritability*) or additive genetic variance (*narrow sense heritability*) to the total phenotypic variance.

Herkogamy: Mechanical obstruction to pollination.

Hermaphrodite: An individual with both male and female reproductive (functional) organs.

Hesperidium: Many-celled fleshy fruit which develops from a polycarpellary superior ovary with axile placentation, the epicarp and mesocarp are fused together forming the rind and endoarp remains thin and papery forming the outer covering of compartments and the edible parts is inner juicy succulent hairs of endocarp.

Heterochromatin: The parts of chromosomes (or chromatin) which take up dark stain during interphase; generally inactive; usually, contains repetitive DNA sequences which are rarely transcribed; may be constitutive or facultative.

Heteroduplex: A DNA molecule in which the two strands are mismatched for one or more nucleotides, *e.g.*, one strand may carry a deletion for some bases.

Heterogenous: A group of individuals (populations) with different genotypes.

Heterokaryon: A mycelium or cell having two nuclei of different genotypes.

Heteroploid: An individual having a chromosome number other than the normal diploid ($2x$) number.

Heteropycnosis: Certain chromosomes or their parts staining differently from the rest of the chromosomes during cell cycle, *e.g.*, in the case of heterochromatin. (*Adj.*, heteropycnotic).

Heterosis: The superiority of a heterozygote over both the concerned homozygotes in respect of one or more characters; in practice, the superiority of an F_1 over both its parents.

Heterostyle: The presence of styles of varying length in different plants of the same species as in brinjal.

Heterothallic: Term used in fungi to designate different mating types (+ & –) of mycelium.

Heterozygote: An individual having two different alleles for one or more genes (*Adj.*, heterozygous; *noun*, heterozygosity).

Heterozygous: An individual having unlike alleles at one or more corresponding loci (opposite of homozygous).

Hexagonal system: A system of planting in which the trees are planted at the corners of an equilateral triangle and thus six trees form a hexagon with the seventh tree at the centre.

Hexaploid: An individual having six genomes of one (*autohexaploid*) or more (*allohexaploid*) kind.

Hfr: In E. coli, high frequency recombination; cells in which the F factor is integrated into the bacterial chromosome.

High density orchard: An orchard where fruit trees are planted at closer spacing to get higher production per unit area without impairing the fruit quality.

High volume spray: In the application of contact weed killers, where through wetting of weeds is done, large volumes (say 50 gallons or more), of liquid is needed. Such sprays are termed as high volume spray.

Hill dropping: A method of sowing in which seeds are dropped at fixed spacing and not at a continuous stream. Thus the spacing between plant to plant in a row is constant.

Hill placement: This refers to applying fertiliser either in bands or near the plants. This method is usually employed when relatively small quantities are to be applied in pinches at the base of each plant of cotton, maize, cauliflower, cabbage, brinjal, tomato and other vegetable crops by this method.

Hill reaction: Photolysis of water by Chloroplasts. This process is involved in photosynthesis. Where the water molecule is split to provide the reducing power necessary for sugar formation.

Hilum: The scar of mark indicating the point of attachment, as in beans.

Histogram: A bar diagrams of a frequency distribution.

Histone: The basic proteins associated with chromatin in eukaryotes.

Holandric Gene: A gene carried in the Y chromosome and, thereby, showing only male-to-male transmission; confined to the male sex only (in mammals).

Holism: An approach that considers all components and aspects of a system; particularly referring to approaches in which material and non-material aspects are considered across disciplinary boundaries.

Home garden: An intensive system of cultivation near the home which incorporates crops for food, medicine and other regular or high value consumption.

Homogametic Sex: The sex which produces only one type of gametes with regard to the sex chromosome.

Homogenous: A group of plants having similar individuals.

Homologous chromosomes: Chromosomes having identical sets of loci. The chromosomes which pair regularly and freely with each other during meiosis; they are identical in gene content and, usually, in morphology.

Homozygote: An individual having two copies of the same allele for one or more genes under consideration (*Adj.*, homozygous; *noun*, homozygosity).

Homozygous: An individual having identical alleles on homologous chromosomes. An organism can be homozygous at one, several or all loci.

Hormone: A compound produced by the cells of one part of the body which acts in the other parts of the organism. Growth-regulating substance occurring naturally in plants or animals; more correctly called growth regulator, not hormone.

Horticultural maturity: The stage of development of a plant or plant part which possess the necessary prerequisite for utilization by consumer for a particular purpose and it depends on the particular crop.

Horticulture: An alternate landuse measure wherein trees/plants that yield fruits, vegetables or flowers are cultivated. This is an ideal method of stabilising the income or resource poor marginal rainfall farmers. Fruit trees increase production and impart a stability to the farming system, especially during the off season. They also provide valuable by-products like fuel wood and leaves and also serve as live fencing. For example: the karonda plant is used as a protective barrier. Horticulture uses 20 per cent of the off-season precipitation which would otherwise be lost as runoff or deep percolation. Dry land horticulture and its ancillary activities like nursery raising, budding, pruning, harvesting, processing and packing can serve as means of providing gainful employment in rainfed areas. The only problem in taking up fruit crops is the longer gestation period and the longer time taken to reach peak production. This problem can be partly be overcome by taking inter-crops during the first few years.

Horticulture, vegetative propagation: Since multiplication of fruit plants can prove to be time consuming, vegetative means are resorted not only to save time but to propagate the desired variety as well. ***Budding*:** a single bud (with or without any wood) is separated from the scion and is applied to the cambium of the stock below the bark. ***Layering*:** a plant twig or shoot is tied down and partially covered so that it can take root while remaining unseparated from its parent. ***Grafting*:** a shoot or a branch is used as a scion and inserted or joining with the root stock. The scion is a short piece or detached shoot containing several dormant buds, which when united with the stock comprises the upper portion of the graft plant. It should be of desired variety and free from disease.

Hortipasture: An arrangement wherein in a field area is used to produce fruit besides serving as a pasture for animals. Fruit trees are raised on the contours while the area between the trees is converted into a pasture.

Host specificity tests: The insects chosen for biological control of weeds should be tested for their basic dietary specifications as there is possibility of an insect adopting new hosts.

Hub crop: It is the crop in sequential cropping system which has the greatest comparative advantage over the crops, *e.g.*, vegetables/truck crops cultivated near city markets.

Hull (rice): The rice caryopsis is surrounded by a hull (husk) composed of two modified leaves, the palea and large lemma.

Huller: A machine used for hulling. While removing the hull, the machine removes part of the bran also.

Hulling: It is the process of removing hull or husk and bran in one operation.

Humid: A climate in which rainfall exceeds potential evaporation during atleast 9 months of the year; refers here to tropical areas which receive more than about 1500 mm annual rainfall.

Humidity: The condition of atmosphere in respect of its water vapour content.

Humification: Process of organic decomposition leading to the formation of humus.

Humus: End product of the degrading process of organic matter that improves soil structure, provides nutrients for plants and increase the capacity of the soil to store nutrients and water.

Hybrid: An individual which is produced as a cross between two genetically unlike parents. The progeny obtained by crossing or mating two individuals or strains having different genotypes.

Hybrid seed: Seed produced by crossing genetically dissimilar plants, *i.e.,* of different varieties and species; the yield potential is superior to that of the parent lines but cannot be maintained in succeeding generations, therefore, the seed must normally be purchased each year.

Hybrid vigour: The unchanged vigour of progeny resulting from the crossing of two different inbred plants.

Hybridization: A method of crop improvement in which two or more plants of unlike genetic constitution in one or more characters and crossed together. This is the process of obtaining progeny through the union between male gametes of one strain with the female gametes of another strain (*Syn.,* crossing, mating).

Hydrograph: A term used to designate materials that absorb moisture from the air, *e.g.,* urea.

Hydrological cycle: The study of incidence and properties of water on and within the ground, including that helds in rivers and lakes. It comprises studies of rainfall, evaporation, run-off, ground water, soil moisture, the hydrological balance, snow and ice accumulation, and chemistry of natural water.

Hydrometer: An instrument used to find the specific gravity of a liquid.

Hydrophilic: Substances showing affinity with water or which are soluble in water.

Hydrophytes: Plants growing habitually in water or in very wet soils where least periodically oxygen becomes deficient as a result of excessive water content.

Hydroponic: The technique of growing plants in aqueous solution of essential nutrients.

Hygroscopic: It is that amount of water which is absorbed by the soil from atmosphere of saturated water vapour due to attractive forces on the surface of the particles.

Hypanthium/Hypanthodium: Type of inflorescence in which the fleshy receptacle forms a hollow cavity with an apical opening guarded by scales, and the flowers are borne on the inner wall of the cavity.

Hypocotyle: Part of the seeding stem below the cotyledons.

Hypogynous: Attachment of other floral parts below the ovary.

I

Identical Twins: Twins derived from the misdivision of a single zygote; members of such twins have the same sex and are identical in genotype. (*Syn.,* monozygotic twins).

Identity: The collective aspect of a set of characteristics of a farm system which are recognised by its users as being in harmony with their culture, their social relations and their relations to crops, animals and nature in general.

Imaginal disc: In insects; a group of embryonic cells determined to develop into specific adult organs; present as thickenings of epidermis in the larvae.

Immature soil: A soil which indistinct or only slightly developed horizons because of the relative short time it has been subjected to the various soil-forming processes. A soil that has not reached equilibrium with its environment.

Immobilisation: The conversion of an element from inorganic to organic combination in microbial or plant tissues. This has the effect of rendering unavailable (and usually and not readily soluble) an element that previously was directly available to plants.

Immunoglobulin: A class of proteins acting as antibodies to specific antigens; insoluble in water but soluble in salt solution; present in blood serum.

Impact evaluation: Impact evaluation refers to assessing the programme impact on the job-performance of the participants.

Impermeable (seed): Impermeable seed is one whose seed coat allows no passage through water or gasses.

Implements: The equipment used on farm to carry out different farm operations either with the help of tractor or bullock power.

Improved seed: The genetically and physically pure seed of an improved seed, breeder's stock seed, foundation seed, registered seed and certified seed.

In-basket exercises: The in-basket exercises are the simulation "That create a complex work organization, rotate participant through key roles in it, and have them "deal with" specific stimulation of a kind they encounter in real life".

Inbred: The progeny of either a single cross-pollination plant obtained by selfing or two closely related plants obtained by inbreeding.

Inbred line: A relatively homozygous line produced through inbreeding and selection.

Inbreeding: Mating between two individuals related by descent, *i.e.*, having a common ancestor.

Incompatibility: (*i*) Inability of the gametes of two organisms to unite and reproduce sexually. (*ii*) Inability of one substance to mix with other without undesirable reaction, *e.g.,* with references to insecticides, herbicides, and fungicides where incompatibility involves loss of efficiency on mixing.

Incomplete dominance: The phenotypic expression of heterozygotes for a gene being intermediate between those of the two concerned homozygotes.

Indehiscent: Describing a fruit or fruiting body that does not open to disperse its contents.

Independent assortment: The segregation of one pair of alleles being independent of that of another pair of alleles, that is, an allele of one gene passes with either one of the two alleles of the other gene with equal frequency.

Indeterminate: Continuing to grow. With inflorescence, the opening of the lower (lateral) flower first, and the terminal ones opening later.

Indeterminate growth: The flowers are borne on lateral branches, the central stem continuing vegetative growth with blooming continued for a long period like in cotton.

Indica rice: All the rice varieties not included in the *javanica* or *japonica* sub-species are grouped together as indicas and are cultivated all over tropical Asia. In this group, the range of variation is very high. They are generally tall, vigorous growing, profuse tillering, late maturing photoperiod sensitive, lodging susceptible and adapted to low fertility conditions.

Indicator: A substance used to test for acidity or alkalinity of a solution by a colour change, *e.g.*, litmus and phenolphthalein.

Indicator plant: A plant which reflects either by its presence or character of growth, specific growing conditions like deficiency of plant nutrient, soil moisture stress, etc.

Indigenous: Describing an organism that is native to an area, occurring or living naturally in a specific area, such as native plants or animals (opposite of exotic); *i.e.*, that has been not introduced from another area.

Indigenous knowledge (IK): Knowledge of the people living in certain area, generated by their own and their ancestors, experience and including knowledge originated from elsewhere which has been internalised by the local people. Also referred to local knowledge.

Induced dormancy: Seed dormancy which is induced by such external factors as unfavourable water supply or temperature though the embryo is quiescent.

Inducer: A small molecule, usually a substrate, which induces the production of an enzyme, usually, concerned with the utilization of that substrate; the inducer combined with the concerned repressor to make the latter inactive and, thereby, relieves the repression of the gene producing this enzyme.

Inert matter: In seed testing, it refers to all foreign matters such as pieces of broken or damaged seeds, empty glumes, lemmas, paleas, soil, sand chaff, stems etc., and all other matter not seeds.

Infiltration: It is the process of water entry into the soil generally (but not necessarily) through the soil surface and vertically downward. Water may enter the soil through the entire surface uniformly as under pending or rain or it may enter the soil through furrows or crevices or move up into the soil from a surface below under high water table.

Inflorescence: The arrangement of flowers on the stem or axis and in relation to each other.

Infrastructure: The supportive features of an economy often provided by government (but sometimes by private industry), *e.g.*, water, transportation, communications, and state organisations.

Inhabitation: Prevention of growth or multiplication of microorganisms.

Inheritance: The reception or acquisition of characters or qualities by transmission from parent to offspring.

Inhibitor: A chemical substance that acts to prevent a process from occurring.

Inhibitory gene: A dominant gene which prevents the expression of another dominant gene; generally, denoted by *I*.

Inoculation: The artificial introduction of microorganisms on or into a medium, living system.

Inorganic nitrogen (Soil): Nitrogen in inorganic form, *viz.*, exchangeable ammonium nitrite, nitrate, fixed ammonium, nitrous oxide and elemental nitrogen.

Input overload: The situation in which the trainers in their eagerness have overloaded the trainees with too much information. As a result the trainees have become exhausted and unable to digest the materials presented and apply them in their work.

Inputs: Refers here to farm inputs, *i.e.*, elements that farmers add to farm resources in order to influence productivity, stability or community. The most elementary farm inputs are water, energy, nutrients and information; examples of other inputs are seeds, agrochemical and equipment.

Insecticide: Any chemical used to control or kill insects.

In situ: In the natural habit.

In situ moisture conservation: A practice adopted for conserving rainwater minimising runoff or percolation to maximise. The objective of this exercise is to preserve maximum moisture in the soil for use by crops. Moisture conservation can be achieved by several methods: Impounding rainwater with contour and vegetative bunds; Adopting appropriate land configurations through tillage; shallow tillage, for instance, aids in water percolation while deep tillage in black soils modifies the soil profile, forming a hard pan which does not allow percolation into the ground water zone; Deep tillage can also be done to break a hard pan as in the case of red soils; Applying mulches during rabi season to reduce surface evaporation. If the crop escapes early floods, then the entire field is harvested about 30 cms above the ground and threshed. The clumps of the late variety continue to grow and if floods do not occur or are not severe, another crop from the late variety is harvested. The main principle here is to have some sort of insulation against nature. In tribal areas, especially under shifting cultivation, seeds of two or more crops are mixed and broadcast with the break of monsoon.

Insolation: The intensity at specific time or the amount in a specified period of direct solar radiation incident that held in rivers and lakes. It comprises studies of rainfall, evaporation, run-off, ground water, soil moisture, the hydrological balance, snow and ice accumulation, and chemistry of natural water.

Instrumented exercises: Instrumented exercises refer to use of psychological instruments such as a rating scale, questionnaire and related forms of data collection techniques, which are used to collect information about trainees' attitude, perception and interest with the purpose of assessing and analyzing their behaviour.

Integrated nutrient management: A practice which involves the judicious use of inorganic, organic and biofertilisers in an integrated manner to maximise production, sustain soil productivity and maintain soil health.

Integrated pest control: Uses a variety of technologies in single pest management with the objective to produce optimum crop yield at a minimum cost taking into consideration ecological and socio-economic constraints under a given agro-ecosystem.

Integrated pest management (IPM): Economic, ecological environmental and social strategy that focus on long-term prevention or suppression of pest problems through a combination of techniques such as encouraging biological control, use of resistant varieties or adoption of alternative cultivation practices or modification of habit to make it incompatible with pest development.

Integrated plant nutrition: Strategy to maintain and possibly increase soil fertility for sustaining crop productivity through optimising all possible sources, organic and inorganic, of plant nutrients needed for crop growth and quality in an integrated way, appropriate to each crop system and specific ecological and socio-economic situation (FAO, 1990).

Integrated weed management: Application of many kinds of weed management technology in a mutually supportive manner.

Intensive cropping: Maximum use of the land by means of frequent succession of harvested crops.

Interactive video: Interactive Video is a very useful training format where programmes are prepared on laser disc.

Intercompatible: The condition when the pollen produced by either variety of a combination is capable of functioning in the styles and fertilizing type ovules of the other variety.

Intercropping: In this system, there is a main crop and a subsidiary crop growing simultaneously. The crops grown in such a system do not compete with each other but are compatible. The period of grain growth of one crop is over before that of the other starts. This practice is followed when the total number of days for a successful crop growth is limited. Sometimes an intercrop of a dissimilar nature is grown as an insurance against total crop failure, particularly in rainfed areas. Mustard are good examples of inter-crops.

Intercross: The crops raised in an orchard or other widely spaced crops for increasing the income from the same piece of land, *e.g.*, short duration vegetables, pulses, oilseeds etc.

Inter-cultivation (inter-cultivating): Soil cultivation performed in standing crop.

Interculture: It is the treatment of soil between rows of crops to remove weeds and to break soil capillaries. Interculture creates a soil or mulch on the soil surface which prevents the cracking of the soil surfaces thus reducing evaporation losses. Interculture is done 3 to 5 times when the crop is on the field depending on the crop duration and the rate of weed infection. Tyned hoes and blade hoes are used for this process.

Interference: Total adverse effect that plants exert on each other when growing in a common ecosystem; includes competition, allelopathy, biotic interference and other modifications detrimental to plant growth.

Inter-fertile: Two or more varieties that are able to cross-pollinate each and produce seed.

Inter-fruitful: The ability of one variety to cross pollinate anther, resulting in either seeded or seedless fruit.

Intergeneric *hybridization*: *Hybridization* between the plants belonging to two different genera.

Intergument: The outer envelope of an ovule.

Intermediate plants: Those plants in which flowering occurs in a small range of intermediate photoperiods but will not flower under light period either shorter than the lower critical value or longer than the higher critical value.

Interminate: Where vegetative growth is not arrested by the first opening of the flowers.

Internal browning of apple: A low temperature disorder characterised by diffuser of brown flesh only in the cortex or the core tissue. It is more evident in a transverse section of the juncture of the stem and core where vascular elements are usually normal and the general texture of the fruit is not noticeably abnormal.

Internal dormancy: Seed dormancy which is internally controlled within the living tissues of seed; also called innate dormancy.

Internal inputs: Inputs which originates from inside the system (farm, village, region, country).

Interphase: The stage in cell cycle when the nucleous of a cell is not undergoing division; a period of active protein (G_1 and G_2 phases) and DNA (S phase) synthesis.

Inter-planting: (*i*) In woodland, the setting out of young trees among existing trees or busy growth; (*ii*) In orchards, the planting of the farm crops along the trees, especially when trees are too small to occupy the land completely; (*iii*) In crop land, the planting of several crops together on the same land like the planting of soybean with maize.

Intersex: A sterile organism whose secondary sex characters are intermediate between males and females of the species.

Interspecific ***hybridization***: *Hybridization* between plants belonging to two different species of a same genus.

Intersterile: Two or more varieties that cannot successfully cross-pollinated and other.

Inter-tilled crop: A crop planted in rows followed by cultivation between the rows.

In vitro: When biological processes are made to occur the organism in a vessel or test outside tube.

In vivo: Within the living organism.

Intervarietal ***hybridization***: *Hybridization* between two varieties of the same kind.

Intraspecific ***hybridization***: *Hybridization* between two genotypes of the same species.

Intron: The parts of an eukaryotic gene which are not represented in the mature mRNA molecules; derived from 'intervening sequences'.

Inventory: It is a list of assets and liabilities which are claims or debts against the business. In other words, it is a detailed list of farm properties, with value assigned.

Inverse nitrogen yield concept: According to Wilcox, the yield of the crop is inversely proportional to its nitrogen, *e.g.*, $Y=K^{-n}$, Where, Y= yield, n = per cent of N in the crop and K= constant (318 lb/acre).

Inversion: Rotation of a segment of a chromosome by 180° so the genes in this segment are present in the reverse order, characteristic inversion loops are produced during meiosis in the inversion heterozygotes.

Inversion temperature: The condition in which cold air settles beneath warm air on a clear night.

Invert emulsion: One in which water is dispersed on oil (instead of oil in water); oil forms the continuous phase with the water dispersed therein.

Involucre: A group of bracts at the base of an inflorescence.

Ionizing radiation: A high energy radiation which produces ions in the medium through which it passes.

Irrigation: The application of water to soil to assist in production of crops. Specially during stress periods. Under irrigated conditions, farming has assured water availability. Rice is grown in about 50 per cent of irrigated area and is the largest consumer of irrigation water in the country.

Irrigation efficiency: It is the ratio expressed in percentage of water stored in the root zone depth of the soil to the water delivered in the field from the farm supply source.

Irrigation period: It is the number of days that can be allowed for applying one irrigation to given area during the peak consumptive use period of the crop being irrigated. It is the basis for designing the capacity of an irrigation project.

Irrigation requirement: It is a part of total water requirement of a crop exclusive of effective rainfall and soil moisture stored in the root zone or that contributed from the shallow ground water table.

Irrigation, drip: Drip irrigation involves the application of water, drop by drop to the root zone of the crop. Water is thus used very economically and the losses due to deep percolation and surface evaporation are minimised. The equipment consists of a pumping unit (to create a pressure of 2.5 kg per sq. cm.), pipelines and drip type nozzles. The amount of water dripping from the nozzle can be regulated to meet the plant requirement. This method is well suited for arid regions and high value crops (fertigation) is now being used to irrigate orchard crops.

Irrigation, pitcher: It involves the application of water either through a crack in a pot or through a wick that is inserted to hole in the pot. The pot is placed deep in the soil or on the surface, close to the base of the plant. The principle is the same as in the drip irrigation.

Irrigation, sprinkler: In this practice, water is sprayed into the air through small nozzles and allowed to fall on the ground at a rate lower than the soil infiltration rate resulting in the use of a very small quantity of water to irrigate a large area. This method is particularly suited to sandy soils with a high infiltration rate, and areas with sloppy topography. The limitation of the sprinkle system is high installation cost.

Irrigation, supplemental: This is the practice of limited irrigation to ensure crop survival in rainfed areas. Supplemental irrigation may also help to sustain the crop during drought periods. Harvested runoff water, which varies between 10 to 40 per cent of total rainfall is stored in a pond or well and recycled into the catchment area. This water is used during the critical periods of crop growth. Supplemental irrigation is also known as life-saving irrigation.

IS elements: DNA segments, of about 800-1400 bp, capable of integrating into, and dissociating from and moving to a new location in the chromosomes of prokaryotes and eukaryotes; they also induce the transposition of other *IS* elements in the same cell.

Isobar: A line of constant barometric tendency. Such lines are drawn on synoptic charts, mainly as an aid to forecasting the movement of features of the pressure distribution.

Isochromosome: A chromosome in which the two arms are identical with each other in gene content and morphology.

Isogenic lines: Lines having identical genotype, except for one gene at which they differ.

Isohyet: A line drawn on a map connecting points with equal rainfall.

Isopleth: A line of equal value of given quantity.

Isotach: A line of equal wind speed.

Isotherm: A line of equal temperature.

J

Japonica rice: These are rice varieties developed in Japan and are adapted for cultivation in the subtropical and warm temperate regions. The Japonica varieties mostly have oval and round grains, the spikelets show distinctive pubsence near the apex and the panicles are compact and dense. Generally they are early maturing having relatively short, sturdy straw, resistant to lodging and respond well to added fertiliser. They are photo insensitive and lack seed dormancy. The rice is somewhat sticky when cooked.

Javanica/bulu rice: Sub-species of *sativa* rice which includes a small number of varities adapted to lower altitudes and are mainly found in Indonesia. Javanica rice is characterized by a straw, long panicle with awned grains, sparse tillering habit, long duration and low sensitive to difference in natural day-length.

Jhum: A complex and comprehensive social system of agriculture used in Nagaland, India; local name for 'slash and burn'.

Jhum cycle: Span of time that is taken to complete the rotation period of the jhumland of a village; time taken from one slash to another.

Job analysis: Job analysis refers to dividing a major work event into its component parts, which allows a trainer to understand what an individual does in the course of his work.

Jointing stage: The stage of growth of cereals following tiller formation. This coincides with the elongation of shoot and thus distinction between nodes and internodes is apparent.

Jonathan spot: A post-harvest disorder of apple, where light brown, irregular or somewhat lobed and slightly sunken to deep brown or black sharply sunken spots at the lenticles are found on Jonathan and number of other cultivars of apple.

June budding: A season of budding exists between late May and early June in the northern hemisphere and November to December in the southern hemisphere.

June drop: The final post-bloom shedding of fruits, often occurring in late May or June.

Juvenile phase: The phase of vegetative growth of a plant before which it cannot be induced to produce flowers. This period may vary from plant to plant, *e.g.*, rice plants have juvenile phase of one month.

K

Kappa particle: In *Paramecium*; a bacterial endosymbiont which harbours a virus; it produces a toxin which kills sensitive paramecia which do not carry the kappa particle; responsible for the 'killer trait'.

Karyotype: The chromosomes of a cell arranged according to their length; also the chromosome constitution of a cell or an individual.

Katabatic wind: Down slope gravitational flow of the colder, denser air beneath the warmer, lighter air results and comprises the Katabatic wind. It is also known as the drainage wind or mountain wind.

Keel: The term used in the papilionaceous flowers where two inner petals are united.

Kernel: Edible seed portion of rice, maize, sorghum etc.

Keylines: Keylines are contour lines without any barriers or earthen bunds on them. They also refer to an unplugged strip or furrow along the contour. Keylines serve as guidelines on a given contour, to take up farming operations and are usually 0.5 metres apart.

Kharif crops: Crop grown during the main monsoon (April to September) such as Kharif sorghum, maize, bajra, rice, cotton, etc.

Kinaesthetic learners: Learners who learn through touch, movements, imitation and practice.

Kinetochore: *See Centromere.*

Klinefelter's syndrome: Sterile human males with the *XXY* chromosome constitution; other associated symptoms as well.

Knot: A speed of one nautical mile per hour and approximately 0.5 metre per second.

Kreb's cycle: A cycle of reaction which plays a central role in the metabolism of many hetrotrophic organism, capable of respiratory mechanism. It involves an enzymes system which converts pyruvic acid to CO_2 in the presence of O_2 with concomitant release of energy in the form of ATP molecules.

Krishi Vigyan Kendras (KVK): The schemes on KVK were launched by the ICAR, Govt. of India, during 1976 as an innovative institution for vocational training in agriculture and allied area. Skill training is imparted to the trainees through the methods of teaching by doing and learning by doing. The KVK aims at practical training to practicing farmers, farm women as well as school drop-outs who acquire training for self-employment.

L

Lab to land programme (LLP): It is an instrument of transfer of technology. This programme was launched in 1979 as a part of the golden jubilee celebration of the ICAR, Govt. of India. Under the project, 50,000 small and marginal farm families and landless agricultural labourers were adapted by Research Institute. Agricultural agencies and voluntary organizations with a view to guide, assist and help them to take advantage of the new technologies developed in the field of agriculture and improve the economic status of the farming lot.

L.D. 50 value: The dose level (lethal dose), which will kill 50 per cent of test animals and is expressed in terms of milligrams of material per kg of animal weight.

Lamella: A structure formed by two membranes lying parallel to each other; usually in form of plates, vesicles or tubles.

Lampbrush chromosome: In diplotene stage of vertebrate oocytes, particularly, in amphibia; the two chromosomes of each bivalent have numerous paired loop-like lateral extensions; loops are the sites of active transcription.

Lanceolate: Wide above the base and tapering to the apex.

Land: The part of the surface of lithosphere which is not usually covered with water, forms the total natural and cultural environment with which production takes place.

Land capability classification: Land capability classification is based on the assumption that every hectare of land should be used in accordance with its capability and limitations. Land is classified according to the most suitable, sustained from erosion or other deterioration. The factors determining land capability classification are major soil characteristics of the land including: texture of top soil and subsoil, slope, extent of erosion, degree of wetness, susceptibility to overflowing and flooding.

Land drainage: The removal of free water both from the land surface and from the soil of the crop root zone.

Land equivalent ratio (LER): It denotes the relatives land area under sole crop required to produce the same yield as obtained under a mixed or an inter-cropping system at the same management level. It is total of the ratio of yield of each component crop in an inter-cropping or mixed cropping system to its corresponding yield when grown as sole crop.

Land forming: Tillage operation which move soil to create desired soil configurations. Farming may be done on a large scale such as contouring or on a small scale, such as ridging or pitting.

Land preparation: It is the breaking open of the land, primarily by ploughing, to make it fit for cultivation. The need for good root system with roots spreading evenly for the absorption of water and nutrients for the plant makes land preparation an essential part of the cultivation process. Land preparations also helps check erosions as it increases the intake rate of water by soil and causes obstruction to surface flow. The objectives of land preparations are: to ensure that the soil has a fine texture, so that there is good seed formation and proper root development; to form an impervious barrier which assists in the retention of water, so that the soil moisture is efficiently used by crops; to destroy weeds. This is especially important for dryland crops where there are too many weeds competing for too little water.

Land reclamation: Making land capable of more intensive use by changing its character and/or environment through such operations as land clearing, controlling erosion, construction or irrigation reservoirs etc.

Land side: That part of the plough that sides along the face of the furrow wall. It helps to slice on the ground and also helps in stabilizing the plough while it is in operation. Land side is fastened to the frog, another part of the plough with the help of bolts.

Land square: An experimental design in which there are same number of rows and columns as treatment, with each treatment occurring once and only once in every row and column, and is suitable when soil fertility gradient in two directions.

Land utilisation index: It is defined as the number of days during which the crops occupy the land during a year divided by 365 and can be expressed as a fraction or as a percentage.

Landslide: Rapid movement down slope of a mass of soil, rock and debris.

Lapse rate: The rate at which temperature decreases in a rising and expanding air parcel.

Latent heat: The quantity of heat absorbed or emitted, without change of temperature, during a change of state of unit mass of a material; it has normally been expressed in the unit cal/gm.

Lateral stream: Lateral streams are water courses or branch drainage lines which lead into the main stream. They are smaller than mainstream and are usually the result of unchecked rill or sheet erosion.

Laterites and Lateritic soils: Laterite is a compact to vesicular rock composed of hydrated oxides of aluminium and iron with small amounts of manganous and titanium oxides etc. They are derived from the atmospheric weathering of several types of rocks under monsoon conditions of alternating wet and dry seasons. The siliceous matter of the rocks is leached away almost completely during weathering. All lateritic soils are poor in lime and magnesia and deficient in nitrogen. Occasionally P_2O_5 may be present in the form of iron phosphate, but K_2O is deficient. There is occasionally a higher content of humus.

Latex: Viscous, excretory fluid which are protein, starch, sugars, resins, alkaloids, enzymes, etc., suspended in a watery or milky fluid and secreted by specialized group of cells or single long cells known as laticifers, secreted by certain plants.

Lath house: A propagation structure made up bamboo stripes which provides outdoor shade and protect container grown plants from high summer temperature and highlight intensities to reduce moisture stress and decrease water requirements of plants.

Lattice design: Any experimental design in which each replication consists of several blocks with randomization occurring within these blocks making possible the comparison of a large number of treatments.

Law of constant return: Constant productivity holds true if all units of the variable factors which are applied to the fixed factor, result in equal additions of the total output.

Law of diminishing return: Diminishing productivity of the variable factors exists when each additional units of input adds less to output than the previous unit.

Lay-by application: Application of herbicides after the last cultivation in field.

Layering: A method of propagation by which a shoot is included to root while still attached to the mother plant.

Leaching of saline solid: If the problem is of salinity, the salts need to be leached below the root zone and not allowed to come up. In practice however, it may be difficult as in the case of deep and fine-textured soils containing more salts in the lower layers. Under these conditions, the provision of sub-surface drainage becomes important. If the soil contains a sandy layer at a lower depth, the leaching salts below this layer will check the rise of salts. Leaching of alkaline soils: For soils with alkaline content soluble calcium salts like gypsum are used to leach it. In course of leaching, the soluble calcium replaces sodium which is leached downwards.

Leaching, nutrient: Process of movement of water soluble chemicals out of soil root zone with the water that drains out of soils.

Leaching, soil: It is the removal of soil material in suspension (or in solution) from a layer or layers of soil. Leaching is one of the methods used to reclaim land. In this process, water is impounded in the field to which chemicals like Gypsum or Calcium chloride or Sulphur are added. Exchange of soluble ions takes place with the desired change required. Simple agronomic techniques like the addition of organic manures, molasses, green manure and crop residues to the soil may also be used for leaching. Leaching is usually done after deep ploughing or trenching, to break the hard pan for improving drainage. It is not effective in cases where the water table is too close to the surface. For example, in coastal areas the upward movement of saline water can cause salt build up in the root zone.

Leaf Area Index (LAI): The total leaf area of a plant or plants divided by the land covered by the plant.

Leaf blade: The expanded flat portion of a leaf.

Learning: Learning is a process by which sensations are perceived, understood, action taken resulting in desirable changes in knowledge, skill and attitude.

Learning style: Learning style refers to different approaches to learning.

Lecture: A lecture consists of oral presentation of information so as to help the listeners to understand the concept, principle and method being presented.

Leeward: Leeward of a point signifies the 'downwind' direction from the point.

Legume: A dry several seeded fruit formed from a single carpel and dehisce on both margins. Any of a family (Leguminosae) of trees, shrubs and herbs (*e.g.*, beans and peas), many of which have the ability to live in symbiotic relationship with rhizoba that can fix atmospheric nitrogen.

Leguminous: Belonging to the same group of plants as pulse crop like beans and peas.

LEIA: Low external input and agricultural based primarily on the use of local inputs (*i.e.*, from the farm and immediate surroundings); makes little use of farm supplies obtained through exchange or purchase.

LEISA: Low-external-input and sustainable agriculture, in which most of the inputs used originate from the own farm, village, region or country, and deliberate action is taken to ensure sustainability.

Lenticel: A usually lens-shaped body of cells formed in the corky layer of bark, which serves as a pore for the exchange of gases between plant and the atmosphere.

Leptotene: The first substage of prophase I of meiosis;

chromosomes clumped as a loose ball of wool, and are visible as fine threads.

Lethal gene: A gene that kills gamete or the zygote bearing it.

Lethal Gene: An allele of a gene which causes the death of all those individuals which carry it in the appropriate genotype; death occurs anytime before the individuals reach adulthood.

Ley farming: This is a crop rotation system in which a grass-legume mixture is grown in rotation with agricultural crops. The system is sometimes called "alternate husbandry" or mixed farming. The grass-legume mixture is called the ley. The grass improves the soil structure and prevents erosion while legume enriches soil nitrogen. Ley farming is economical for rainfed farmers as they need not invest much in nitrogen fertiliser input for food grain crops but can depend on the nitrogen built up by pasture legumes. In rainfed farming, it has been observed that stylosanthes hamata raised for a period of 2-3 years in a 4 yearly period would help in building soil nitrogen to the extent of 35 kg per hectare. With this a farmer need not use any nitrogen for the subsequent sorghum crop.

Lianes: Vines with woody stems, common in forests of warm climate.

Ligase: An enzyme which seals the nicks or the cuts in a DNA strand.

Lightning: A visible electric discharge (lightning flash) associated with a thunderstorm.

Line sowing: It is the sowing of seed in a line or row with optimum distance maintained between rows and inter-row seed furrows. Line sowing, though laborious when compared to broadcasting, has several advantages. It helps in ploughing and weeding between rows, facilitating even the use of mechanical weeders. It helps in the even distribution of fertilisers. Recommended spacing for each crop or variety can be obtained from the local research stations.

Linkage: The tendency of genes to stay together during inheritance; due to the genes being located relatively close to each other in the same chromosome; produces typical distortion of test cross ratios (of F_2 ratios as well).

Linkage map: A linear map of the genes showing linkage with each other; it depicts the sequences in which genes are located in the chromosome as well as the frequency of recombination between the adjacent genes.

Lipophilic: Substances showing affinity with fats like Easters or which are soluble in fats.

Litter: Uppermost layer of organic material in the soil surface, including leaves, twigs and flowers, freshly fallen or slightly decomposed.

Livelihood system: A combination of people, resources and environment in which the stock and flows of food and cash are used to meet the basic need of people. The livelihood system of rural household may include cropping, tree growing, animal keeping, fishing, hunting, gathering, processing, trading, paid employment and a wide variety of other non-farm activities.

Local experts: A group of knowledgeable and respected men, drawn from various tribe groups, who assist in the planning, implementation and monitoring of rural development activities.

Localized incompatibility: Cases of graft incompatibility in which the incompatibility reaction can be overcome by insertion of a mutually compatible interstock piece between the stock and scion.

Locule: A compartment of an ovary, anther or fruit.

Locus: The fixed position in a chromosome occupied by a gene; often used as a synonym of gene since, ordinarily, genes occupy a fixed position in the concerned chromosome.

Logarithmic sprayer: A device for spraying a pesticide at a steadily decreasing rate of application.

Long day (LD) plant: One that is stimulated by long days to initiate some process usually flowering.

Long-wave radiation: The radiation which is emitted by earth and atmosphere. The radiation is confined within the wavelengths of about 3 to 100 km.

Lopping: Removal of unwanted branches or twigs.

Low volume spray: In translocated weedicides complete wetting of the foliage is not necessary and thus smaller quantities of liquid (say 5-20 gallons per acre) is needed. Such sprays are termed as low volume spray.

Lysimeter: An instrument for measuring the rate of percolation of rain through a soil.

Lysis: The rupture of a bacterial cell due to the disintegration of its cellular structures following its infection by a virus. (That is why bacterial viruses are called bacteriophage.)

Lysogeny: The chromosome of a bacteriophage becoming integrated into the chromosome of its host; in case of temperate phages; such bacteria are called *lysogenic.*

M

M_1, M_2: The first and second generations after mutagen treatment.

Macronutrients: These chemical elements are required in large quantities for the healthy growth of plant. They are applied in the form of fertilisers or liming material. The macronutrients are Nitrogen, Phosphorous, Potassium, Calcium, Sulphur and Magnesium.

Main stream: This is the main water course that runs through the watershed. It is fed by the lateral streams or the branch drainage lines from the higher reaches. The main stream is the catchment drain for the entire watershed.

Maintenance training: Training provided to impart specialized subject matter knowledge to the employees so that they remain updated with regard to latest developments in the field.

Major modes of disease transmission: Seed-borne diseases: Here pathogens are carried either on the surface of a seed or within it. Externally-borne diseases are controlled by pre-treating seeds with chemicals. Control of internally borne diseases requires hot water treatment and sun-drying. For instance, wheat seeds infected with loose smut can be soaked for 4-5 hrs in ordinary water. They are then immersed in hot water at 490°C for one minute. Finally, they are immersed in hot water at 140°C for 10 minutes to disinfect them. For solar treatment, the seed is soaked for 4 hours and then spread in a thin layer to dry it in the sun. Soaking causes the seed to germinate and sun drying kills the pathogens.

Male sterility: The male gametes produced by an individual being non-functional.

Manure: Solid or liquid animal waste often mixed with organic waste and applied to soil.

Manuring: Application of animal dung, compost or other organic material used to fertilise the soil.

Map unit: *See, Crossover Unit.*

Marcotting: The practice of binding moist moss onto branches to promote new root which may be served from the mother plant and planted elsewhere.

Marginal land: It is land that gives returns barely enough to cover the cost of a specific use. Most marginal lands are very low in productivity. Food and fodder cannot be raised on them economically. The lack of vegetative cover on most of these lands is the cause of serious soil and water erosion. About 107 Mha our land is in some state of degradation. If this is not controlled, precious arable land will deteriorate into marginal lands.

Mass selection: Selection of phenotypically similar plants and the next generation propagated from the aggregate of their seed.

Maternal effect: The phenotypes of progeny being affected by the genotypes of their mothers.

Maternal inheritance: The transmission to progeny of a character from the female parent only; *see, Cytoplasmic Inheritance.*

Mattocking: The practice of removing banana pseudostem after harvesting of fruit.

Mealy breakdown: A disorder of apple characterised by the whole fruit becoming dry and mealy with the skin rupturing and the flesh breaking open which is associated with aging and is aggravated by advanced maturity, delayed cooling, high storage temperature, high relative humidity, extended storage, large fruit and occurrence of water core.

Mean: The sum of all the observations in a sample divided by the number of observations in the sample. (*Syn.*, average).

Mean sea level: Mean plain at which the tide oscillates the average height of the sea for all stages of the tide. At any particular place, it is derived by averaging the hourly tide heights over a 19 years period.

Mechanical dormancy: Type of seed coat dormancy in which seed covering are too hard to allow the embryo to expand during germination.

Mechanism of action: Precise biochemical or biophysical reaction or series of reactions that culminate into the final or ultimate effect of a herbicide; many herbicides have a primary and secondary mechanism of action.

Megaspore: A spore having property of giving rise to a female gamete.

Megasporogenesis: The production of female spore through meiosis in the ovules of plants.

Meiosis: Two cell divisions, occurring one after the other, following only one DNA replication; shows chromosome pairing and chromosome separation in the first division, and chromatic separation in the second; leads to the formation of haploid daughter cells (spores in plants; gametes in animals).

Meiotic drive: A mechanism that leads to the production of unequal frequencies of functional *gametes* by a heterozygote.

Melting point: That temperature, characteristic of given substance at a given pressure, at which the change of state from solid to liquid occurs.

Mendelian population: A random mating population; a group of individuals sharing the same gene pool.

Merclinial chimeras: A type of chimera in which the mutated tissues, occupy only a part of one of the layers and does not extend completely around the shoot.

Meri-cloning: Vegetative multiplication through meristem culture.

Meridional flow: Air flow in the direction of the geographical meridian, *i.e.*, south-north or north-south flow.

Meristem: A tissue primarily concerned with protoplasmic synthesis and formation of new cells by division.

Meristem culture: A type of *in vitro* culture technique where shoot or apical meristem along with some surrounding tissue is grown on suitable culture media which is employed for clonal propagation and recovery of virus free plants.

Merozygote: In bacteria; a partial zygote; a cell having a segment of DNA (*exogenote*) homologous to a segment of its own chromosome (*endogenote*).

Mesocarp: The middle layer of the fruit between epicarp and endocarp such as pulp and flesh of different fruits.

Mesocotyl: The elongated portion of the exis between the cotyledon and the coleoptile of a grass seedling.

Messenger RNA (mRNA): The RNA molecules which carry the information for amino acid sequences of specific proteins which is deciphered with the help of ribosomes and specific tRNAs.

Metabolism: The sum total of all biochemical processes concerned with energy production and energy utilization occurring in living cells.

Metacentric Chromosome: A chromosome with the centromere in its centre or close to the centre; the two arms of such a chromosome are nearly equal.

Metafemale: A female individual, usually sterile and weak, showing a more intense expression of the secondary sex characters; due to a greater number of *X* chromosomes (*e.g.*, *XXX*) than in the normal females (*XX*) (*Syn.*, superfemale).

Metaphase: The stage during cell division after prophase and before anaphase; the chromosomes are short, thick distinct from each other and arranged on the equatorial plate.

Metaxenia: The effect of pollen on the development of material tissues of the fruit.

Micro climate: Climate of a small area such as surrounding an individual plant.

Micro nutrients: Chemical elements required in extremely small quantities are called micro nutrients. These are Boron, Chlorine, Copper, Iron, Manganese, Molybdenum and Zinc.

Microbial insecticide: A class of insecticides produced from naturally occurring micro-organisms, *e.g.*, *Bacillus thuringiensis.*

Microbudding: A method of budding in which the bud piece is reduced to vary small size, but it is not done in aseptic conditions.

Microclimate: The physical state of the atmosphere close to a very small area of the earth's surface. Often in relation to living matter such as crops or insects. The temperature, sunlight, humidity and other climatic conditions in a small localised area, *e.g.*, in a field, in a stand of trees, in the vicinity of a given plant, or in the topsoil.

Microlab: Microlab consists of a number of mini exercises to help the partcipants get prepared and receive a maximum output from the programme.

Microspore: One of the four cells produced by the meiotic division of the microspore mother cell.

Microsporogenesis: The production of haploid male spores (pollen grains) through meiosis in the anthers of plants.

Microtubules: The hollow, tube-like filaments formed by precisely arranged protein molecules; present as constituents of many cellular structures, *e.g.*, centrioles, flagella, spindle fibres etc.

Millerandage: A disorder of grape, in which the ovary persists but the seed remain small or do not attain the usual size.

Mind map: A creative sequencing technique that can be used successfully for content sequencing.

Minimum temperature: The lowest temperature attained, usually in the thermometer screen or on the ground, during a period of time.

Minimum tillage: Soil management practices which seek to minimise labour inputs and soil erosion, to maintain soil moisture and to reduce soil disturbance and exposure. Crop stubble is left or mulch is applied to protect the soil. Also known as conservation tillage or reduced tillage. In its most extreme form (zero-or-no tillage), seeds are drilled directly into the otherwise undisturbed soil.

Miscible: Two or more liquids which, when combined together, form a uniform stable mixture.

Mist: A state of atmospheric obscurity produced by suspended microscopic water or droplets or wet hygroscopic particles. The term used for synoptic purpose when visibility is equal to or exceeds 1 km. The correspondence relative humidity is greater than 95 per cent.

Mitochondria: Cytoplasmic organelles enclosed by two concentric membranes, the inner membranes having characteristic infolds called cristae; concerned with oxidative phosphorylation and production of ATP; contains DNA and participates in cytoplasmic inheritance.

Mitosis: Equational cell division, as in somatic cells, in which the number of chromosomes is not reduced. Each nuclear division is preceded by one round of DNA replication; during anaphase sister chromatics separate and pass to the opposite poles.

Mitra kisan: A Mitra kisan or contact farmer is a link between the scientists and farmers. This is a concept under the NWDPRA watersheds.

Mixed Cropping: In this system, two different crops, or two varieties of the same crop with different durations, are grown together simultaneously on the same piece of land, without any definite row arrangement. For example, in high rainfall areas, prone to flood damage late in the crop season, an early variety of rice is grown along with a long duration rice variety in the proportion of 1:3 or 1:4.

Mixoploidy: Chimera having different ploidy levels in the L1, L2, L3 and L4 layers of apical meristem.

Mode of action: The sequence of events that occur from an herbicide's first contact with a plant until its final effect (often plant death) is expressed.

Modified leader: A training method in which the main trunk is allowed to grow to about 1.8-2.5m in height and then cut back for giving a shape intermediate between central leader and open centre.

Modifier genes: Genes that affect the expression of non-allelic gene(s).

Modifying gene: A gene that increases the phenotypic expression of a major gene; usually, several modifying genes act in an additive manner and generate a continuous variation in an otherwise qualitative trait.

Moist air: A term which usually signifies simply a mixture of dry and water vapour. In synoptic meteorology and climatology the term is applied to air of high relative humidity.

Moisture conservation project: Since water is scarce and precious in rainfed areas, it is essential to conserve the same. Land treatment should incorporate specific practices for the conservation of water. Possible methods include: In situ moisture conservation; Recycling; Water harvesting; Mulching; Tillage-off-season, deep and shallow Interculture; Vegetative cover and barriers; Land configurations.

Moisture stress: The tension at which water is held by the soil.

Monadelphous: Having the filaments united into a single tube around the gynoecium.

Monitoring: Monitoring refers to 'continuous or periodic review and surveillance (overseeing) by management at every level of the hierarchy of the implementation of an actively to ensure that input deliveries, work schedules, targeted outputs and other required actions are proceeding according to plan.'

Monocotyledon (Monocot): Any seed plant having a single cotyledon or seed leaf.

Monocropping: Mono cropping refers to growing of only one crop in a crop growing season. Repeated growing of the same sole crop on the same land.

Monoculture: Growing of a single species of plant in a given unit area.

Monoecious: A trait of having separate male and female flowers on the same plant.

Monoecious: In plants; a plant having both male (*staminate*) and female (*pistillate*) flowers.

Monogenic: Governed by or involving a single gene.

Monohybrid: The progeny derived by mating or crossing two individuals or strains which differ for the gene (in practice, one character).

Monoploid (x)**:** An individual having one copy of a single genome; haploid from a diploid ($2x$) species.

Monopodial: Axis produced and increasing in length by apical growth.

Monosomic: An individual having one chromosome less than the normal somatic complement of the species; denoted by $2n$–1. An individual in which one member of a homologous pair is missing.

Monotypic: A genus with a single species.

Monozygotic twins: *See, Indentical Twins.*

Monsoon: Derived from the Arabic word 'mausim' meaning 'season' the term originally referred to the winds of Arabian sea which blow for about six months from the south-west. But is now used also for other markedly seasonal winds.

Mosaic: An individual or tissue having cells of two or more different genotypes.

Mother plant: A seed bearer; a plant from which vegetative portions have been selected for propagation.

Mulch: Organic material, such as peat, leaf mould, or shredded bark, that is spread out on ground to suppress annual weeds; also applied around the base of tree and shrubs to help absorb and retain water and add nutrients. It also prevents evaporation of moisture, regulate temperature and control weeds.

Mulching: A practice by which materials such as straw, saw dust, manures, plastic film, vegetative wastes of crop residues are spread upon the surface of the soil. The practice protects the soil from surface evaporation and effects of splashing rain drops, soil crushing and freezing. The cover used is called the mulch. There are different kinds of mulches. Surface mulches are used to: Prevent soil from being blown or washed away; Reduce evaporation; Increase filtration; Increase crop yields. Some commonly used mulches are straw, leaves, and green manure crops. Soil mulches are a thin layer of loose soil that can be created by frequently stirring soil with surface tillage implements like a blade harrow. Soil mulching prevents deep cracking of soil.

Multiple alleles: More than two alleles of a single gene (more precisely, of a single *cistron*).

Multiple cropping: Growing two or more crop in the same field in a same year, at the same time or one after the other, or a combination of both.

Multiple cross: A cross involving more than two parents.

Multiple factor: The type of inheritance in which the character is dependent on many genes of factors, the influence of environment on such characters is large.

Multiple fruit: A collection of fruits which develops by fusion of carpels from many flowers (an inflorescence) plus stem axis and accessory tissue.

Mutable gene: A gene having an unusually high spontaneous mutation rate.

Mutagen: A chemical or physical agent which induces mutation.

Mutant: An organism or cell showing a mutant phenotype due to the mutant allele of a gene.

Mutation: A sudden heritable change in a gene or chromosomal structure which is not due to segregation or recombination.

Mutation, gene: A change in the base sequence of a gene, usually producing a mutant phenotype.

Mutator gene: A gene which causes another gene or genes to undergo spontaneous mutation.

Muton: The unit of mutation; the smallest unit of a gene capable of undergoing mutation; represented by one nucleotide.

Mycorrhiza: Symbiotic associations of the threadlike filaments of a fungus with the roots of higher plants, which can increase the plants capacity to absorb nutrients from the soil.

N

Nagamese: A local dialect based on Assamese, Bengali and Hindi commonly used in Nagaland.

Naked chilling: Stratification of seeds in plastic bags without a surrounding medium.

National Agricultural Research Project (NARP): It is a project launched by the ICAR, Govt. of India, to strengthen the regional research capabilities of the agricultural universities in India, with an objective of conducting location-specific and production oriented research in all the agriclimatic zones identified in the area of its jurisdiction.

National demonstrations: The national demonstration project was launched by the ICAR in 1967 as an instrument of transfer of technology. The main objectives are: (i) demonstrating to the farmers the production potential of land by adopting high yielding varieties, multiple-cropping programmes and other scientific package of practices. (ii) Providing research workers a first-hand information on problems faced by farmers and factors limiting crop yield.

Natural farming: A system of agricultural devised by Masanobu Fukuoka that seeks to follow Nature by minimising human interference: no chemical cultivation, no synthetic fertilisers or prepared compost, no weeding by tillage or herbicides, no dependence on chemicals.

Natural selection: Differential rate of reproduction of different genotypes of an organism in response to the environmental factors; the basis of organic evolution.

Natural voice: A voice, which reflects the trainers' true personality and sincerity. It creates a congenial and friendly atmosphere.

NDFF (Nitrogen derived from Fertiliser) per cent: Percentage of N in the plant derived from the fertilizer applied and it is calculated by using the formula: per cent Ndff = per cent ^{15}N excess in sample/per cent ^{15}N excess in fertiliser × 100.

Necrosis: The death of plant tissue.

Nematodes: Eelworms, found in great quantities in moist topsoil, many of which are parasitic on plants and animals.

NEPED: Nagaland Environment Protection and Economic Development (NEPED) project through people's action; project supported by the Government of Nagaland, the India-Canada Environment Facility (ICEF) and the International Development Research Centre (IDRC).

Net area sown: Class of land under revenue classification represents actual extent of land sown to crops during a given year, when cropped more than once in any year, its area is added to the net area sown to arrive at the total area sown during the year.

Net photosynthesis: When photosynthesis is measured with correction for respiration.

Net productivity: The arithmetic differences between calories produced in photosynthesis.

Net radiation: The net radiation is the differences between total upward and downward flux and is a measure of the energy available at the ground level.

Networking: Establishing and strengthening lengths between individuals, groups and organisations with similar interest and objectives.

Neutrophiles: The plants which can grow in neutral soils (*p*H range 6.5-7.4).

NGO: Non-governmental organisation; non-profit, voluntary group engaged in relief and/or development activities.

Niche: A space in the ecosystem which, because of the specific local ecological, physical and/or social characteristics, is suited to a particular plant or animal species or a particular activity.

Nicotine: An alkaloid produced as a secondary plant product, by trichomes (glandular hair present on the leaf surface) as secretion in tobacco, which has some insecticidal properties especially against aphid (*Myzus persicae*).

Night soil: Night soil is human excrement, solid and liquid. In India, it is directly applied to the soil to a limited extent and as town compost to a large extent. In cities, which have sewage facilities, sewage water and sludge are used directly to raise crops. On an average night-soil contains 5.5 per cent nitrogen 4.0 per cent phosphorus (P_2O_5) and 2.0 per cent potash (K_2O) on oven basis.

Nimbostratus (Ns): Grey cloud layer, often dark, the appearances of which is rendered diffuse by more or less continuously falling rain or snow, which is in most cases reaches the ground.

Nitrate of soda and potash: Chiefly the sodium and potassium salt of nitric acid containing not less than 15 per cent of nitrate nitrogen and 10 per cent of potash (as K_2O).

Nitrate reductase (NR): A metalleflavoprotein enzyme that catalyses the reduction of nitrate to nitrite.

Nitrate reduction: Reduction of nitrate to the level of ammonia with the help of enzymes.

Nitrification: It is a two-step process in which the ammonia or ammonium is converted to nitrite and finally to nitrate and these two steps are carried out by the autotrophic microorganisms Nitrosomonas and Nitrobacter respectively.

Nitrification inhibitors: Chemicals used to control nitrogen transformations in the soil. Due to their texic effect on nitrifying bacteria, they temporarily inhibit the production of nitrate, *e.g.*, N-server, nitrapyrin, potassium azide etc. Application rate 0.2 to 0.6 kg/ha of active ingredient.

Nitrogen assimilation: The incorporation of nitrogen compounds into cell substances by living organisms.

Nitrogen fixation: The conversion of elemental nitrogen to organic combinations, or to forms readily utilizable in biological processes, by nitrogen-fixing micro-organisms. When brought about by bacteria in the root nodules of leguminous plants, it is referred to as symbiotic, if by free-living micro-organisms acting independently, it is referred to as non-symbiotic.

Nitrogen fixing plant: A plant which can assimilate and fix the free nitrogen of the atmosphere by the aid of bacteria living in the root nodules. Legumes with the associated rhizobium bacteria in the root nodules are the most important nitrogen fixing plants.

Nitrogen free extract (NFE): It consists mainly of carbohydrates like sugars and starches along with small amounts of pentesans and other less important carbohydrates NEE/100 = (Moisture +crude fibre).

Nitrogen-fixing: The ability of organisms (bacteria, actinomycetes or algae) to convert atmospheric nitrogen into a form, which can be used by higher plants. These organisms may be free-living or live in a symbiotic relationship with the plants.

Nitrophosphates: Products obtained by treatment of phosphate rock with nitric acid alone or in admixture with sulphuric or phosphoric acid, without subsequent treatment with ammonia.

Node: A point on the stem from where the leaf/flower arises.

Non-arable land: The land which is considered unfit for growing agricultural crops. They occupy classes V to VIII under the land capability classification. These lands are suitable for perennial vegetation including grasslands, forests or horticulture and for recreational purposes. Non-arable land normally constitutes an integral part of the watershed area. The productivity of cultivable land depends to a very large extent on the treatment of non-arable lands. The objectives of treating non-arable lands are: to control runoff, increasing the moisture regimes of the soil and reducing soil erosion; to recharge ground water; to increase productivity; to maintain the ecological balance; to create self-sustaining systems through the development of non-arable land.

Non-acid forming fertilizer: One that is not capable of increasing the residual acidity of the soil.

Non-climacteric fruits: Fruits showing a gradual decline in respiratory rate with ripening.

Non-disjunction: The failure of a chromosome pair (during A I of meiosis) or of sister chromatids of a chromosome (during A II of meiosis and anaphase of mitosis) to separate during anaphase; as a result, they pass to the same pole leading to the production of $n + 1$ and $n - 1$ gametes (meiosis) or $2n + 1$ and $2n - 1$ daughter cells (mitosis).

Non-formal training: Training conducted based on non-formal modalities such as participatory action research where there is flexibility with regard to choice of training contents, methodology and evaluation.

Non-polar compound: they are relatively uncharged molecules held together by van der Wall's forces and usually exhibit low water solubility and high oil solubility.

Non-recurrent: Term used in back-crossing where a particular parent is used only once (*q.v.* donor parent of recurrent parent).

Non-recurrent apomixis: A type of apomixis where embryo arises directly from the haploid egg nucleus without fertilization.

Non-renewable resources: Resources such as oil, coal, mineral ores which cannot be naturally regenerated on a time scale that is relevant to human exploitation (IIED 1988).

Non-saline-alkali soil: A soil that contains sufficient exchangeable sodium to interfere with the growth of most crop plants and does not contain appreciable quantities of soluble salts. The exchangeable sodium-percentage is greater than 15 and the electrical conductivity of the saturation extract is less than 4 milli mhos per centimetre (at 25°C). The *p*H of the saturated soil paste is usually greater than 8.5.

Non-selective herbicide: A material that tends to kill all plants with which it comes in contact.

Non-symbiotic nitrogen fixation: Fixation of molecular nitrogen by free-living microorganisms like clostridum and Azatobactor.

Nor'wester: A violent, convective type of storm, often accompanied by alines squall, which occurs in Bengal and Assam in the months of March to May. The storms are also named because of their pronounced tendency to move from the north-west.

No-tillage planting: Direct planting in essentially unprepared seed beds.

Noxious weed: A weed plant especially undesirable, troublesome and difficult to control, *e.g.*, *Cyndon dactylon, Cyperus retundus, Sorghum balepense,* etc.

Nucleic acid: The macromolecules composed of phosphate, pentose and organic bases, *viz.*, DNA and RNA.

Nucleolus: A relatively dense, spherical structure rich in RNA and protein, and associated with the NOR; the site of ribosome assembly and synthesis of ribosomal components.

Nucleolus Organiser Region (NOR): The segment of a chromosome containing the genes concerned with the synthesis of ribosomes; usually, seen as a constriction in metaphase chromosomes and associated with the nucleolus during the prophase I of meiosis.

Nucleoprotein: A complex containing nucleic acid and protein, *e.g.*, chromatin.

Nucleosome: The bead like structure of chromatin fibres; each composed of a histone octamer and 146 base pairs of DNA.

Nucleotide: The unit of DNA and RNA organisation; each nucleotide has one molecule each of phosphate, pentose (ribose in RNA, deoxyribose in DNA) and one of the five organic bases [A, G, C, T in DNA) and U (in the place of T in RNA)].

Nucleus: In eukaryotes; the organelle containing chromosomes and nucleolus, and surrounded by a typical nuclear envelope composed of two concentric membranes.

Nullisomic: An individual having on chromosome pair less than the normal somatic complement of the species.

Nullisomic: An individual in which a homologous pair of chromosomes is missing.

Numerical weather forecast: An objective forecast, in which the future state of the atmosphere is determined by the numerical solution of the basic theoretical equations involved. The calculations are so lengthy as to necessitate, in general, the use of the electrical computer.

Nurse crop: A companion crop which nourishes the main crop by way of nitrogen fixation and/or adding the organic matter into the soil, *e.g.*, cowpea intercropped with cereals or new plantations of fruit trees.

Nurse root grafting: A temporary graft used to induce the scion to develop its own roots. The nurse root sustains the plant until the scion develops its own root.

Nursery: It is a store-house for all planting materials that may be needed for any land development activity. A nursery contains different species of trees, shrubs, grasses and horticultural plants suitable for a particular agro-climatic zone. The presence of a nursery is critical to a watershed area, especially for the treatment of unproductive and eroded land. There are two types of nurseries typically prevalent in a watershed area, mother nurseries and kisan nurseries.

Nursery, kisan: Farmers within a watershed area are encouraged to start nurseries with plants and grasses of their choice. They are given monetary incentives. Since irrigation is a critical factor, farmers with irrigation sources are given first preference. The planting material is given to the farmer at a nominal rate and the farmer receives seeds, seedlings, polythene bags, manure bags and the technical know-how for maintaining the saplings. The saplings are maintained by the farmers till they are ready for planting.

Nursery, mother: It is a composite nursery run by the watershed task force to cater to the entire watershed and even the adjoining areas, at times. It contains all the planting material required for afforestation, horticulture, seeds for pasture plants for vegetative barriers and other useful plants needed by the community. The planting materials are taken from the mother nursery by the farmers either directly for their fields or to their own (kisan) nurseries.

Nursery school: In order to inculcate interest and awareness about trees and their importance among school children, school nurseries are taken up by school authorities with assistance from the agricultural department. The children are thus familiarised with the various watershed concepts at an early stage.

Nut: A hard, indehiscent, usually one-seeded fruit, derived from carpels plus stem axis plus accessory parts.

Nutirents: These are elements needed for plant metabolism. Atleast 16 plant food elements are necessary for the growth of green plants and in the absence of anyone of these, a plant fails to complete its life cycle. Of the 16 elements, Carbon, Hydrogen, Oxygen and Nitrogen are very essential for plants and are also called primary nutrients. Plants obtain carbon from carbon dioxide in the air and oxygen and hydrogen from water. The remaining elements are drawn from the soil. Based on the relative amounts normally found in plants these nutrients are further classified as macro nutrients (if large amounts are involved) and micro nutrients (if only traces are involved).

Nutrient: Substance, which is used by an organism for production of energy, maintenance and production of body and regulation of life-sustaining process.

Nutrient cycle: Path of a nutrient through the ecosystem. Including its assimilation by organisms and its regeneration in a usable inorganic form.

Nutrient cycling: The pathway of nutrient substances from occurrence in the physical environment to their incorporation in the living organisms and their return to the physical environment through the metabolic activity, death and decay of organisms.

Nutrient harvesting: Deliberate activities to capture nutrients from outside the farm system or from other parts within the farm system and to concentrate them in particular area in the farm.

Nutritive values: Nutritive value of forage is characterised by the chemical composition, digestibility and nature of the digested products. Chemical composition is a factor associated with only the plant and its environment, but digestibility, the nature of digested products and their efficiency of utilization are associated with both the plant and animals.

O

Obligate parasite: A weed never found in the wild stage, but grows only in association with another plant, as a parasite.

Octaploid: A cell or an organism having eight identical (*autooctaploid*) or different (*allooctaploid*) genomes (8 X chromosomes).

Offset: A short shoot that rises from an auxiliary bud near the base of the stem and gives rise to a daughter plant at its apex.

Off-type: Plants of seeds deviating significantly from the characteristics of a variety as described by the breeder in any observable respect.

Oil (plant): A slippery viscous liquid or fat like substances produced and stored in the seeds species such as sunflower, groundnut, soybean etc.

Oil cakes: Oil cakes are organic manures that are the residues of the oil industry. They have a high nutrient content. It is the by-product of the oil extraction from castor, sunflower, coconut, groundnut, mustard, linseed, niger, sesamum and neem among others. Oil cakes are of two kinds: Edible oil cakes that come from the produce of coconut, groundnut, sesamum, rapeseed, mustard, linseed, niger and cotton seed after the oil has been extracted. Most of these cakes are used as animal feed and manure. Non-edible cakes come from castor, sunflower and neem. These are used only as manure.

Okta: Unit, equal to area of one-eighth of the sky, used in specifying cloud amount.

Oligogene: A gene producing a marked effect on the expression of a character so that the individuals can be easily grouped into separate classes on the basis of this trait; its expression is relatively little affected by the environment (*cf.* polygenes).

Oncogene: A gene which transforms an animal cell into a tumour forming or cancerous cell; *cellular oncogenes* are an integral part of the animal genome, while *viral oncogenes* are present in the genomes of tumour producing viruses; both are very similar.

One year seeding, seven years weeding: The explanation of this proverb is that if we allow the weeds to produce and shed their seeds in the field, then the weeds cannot be removed from the field in spite of our repeated weedings even for 7 years due to their enormous seed producing capacity.

On-the-job training: On-the-job training takes place in the work place and consists of helping an individual trainee to acquire new skills and expertise through practicing them under the guidance and supervision of an experienced trainer or an employee.

Oocyte: The diploid cell in the ovary of a female animal which undergoes meiosis to yield the ovum or egg cell.

Oogenesis: The formation of eggs or ova in female animals through meiosis in oocytes.

Open learning: Open learning is an approach, which takes place in a way, and at a time, place and pace, convenient to the learner.

Open-ended questions: Open-ended questions are such questions for which there are several answers.

Open-formula mixed fertiliser: Fertiliser mixtures of which the composition and the complete make-up are disclosed by manufactures. Thus the fertiliser grade, the straight fertilisers and the filler used are disclosed on each bag.

Open-pollination: Pollination without control.

Open well: These are wells dug down to the water bearing strata. When the water table is shallow—a circular, rectangular or square hole (generally three feet or more in width) is dug to reach the water table. Water driven upto the surface is used either for irrigation or for drinking purposes.

Operational research project (ORP): It is an instrument of transfer of new technology. It is an integrated approach launched by ICAR in co-operation with local agencies, voluntary organisation, agricultural universities, and state agricultural departments to test, adopt and make them suitable for large scale adoption by farmers, to find out the constraints that impede acceptance and release of such findings and to find out the profitability of such findings.

Operator gene: In operons; the gene to which the repressor molecule binds; it is the gene adjacent to the first structural gene of an operon.

Operon: A group of structural genes whose transcription is regulated by a set of regulator, promoter and operator genes, which are often located next to (upstream or onto the left of) the structural genes.

Opportunity cropping: The practice of placing primary emphasis on the use of stored soil moisture while determining or not to establish a crop.

Optimum: The best or most favourable condition, degree or amount for a particular situation.

Optimum-economic dose: If there is a quadratic response between input (X) and output (Y) $Y = a + bx + cx^2$, the optimum economic dose is given by $X_{opt} = q/p/2c$, where, q = cost of unit x, p = price of unit y, b & c are constants.

Organelle: A specialized substructure of a cell with a specific structure and function.

Organic: Any chemical compound containing carbon or derived from living organisms.

Organic farming (biological husbandry): It is agricultural production system which avoids or largely excludes the use of synthetically compounded fertilisers, pesticides, growth regulators and livestock feed additives. To the maximum extent feasible organic farming system rely upon crop rotation, crop residues, animal manure, legumes, green manure, mineral-bearing rocks and aspects of biological pest control to maintain soil productivity and tilth to supply plant nutrients and to control insects weeds and other pests.

Organic nitrogen (soil): Nitrogen in the form of organic compounds, *e.g.*, amine acid, amine suger, purine and pyrimidine bases.

Organic nitrogen fertiliser: Fertiliser containing nitrogen associated with carbon in organic combination.

Organic phosphorus: Phosphorus present as a constituent of an organic compound or a group of organic compounds, *e.g.*, glycero-phosphoric acid, inesital phosphoric acid, cytidylic acid.

Organic soil: Soils containing organic matter in sufficient quantities to dominate the soil characteristics. Frequently all soils containing 20 per cent or more organic matter by weight is arbitrarily designated as organic soil.

Orientation or induction training: This training is given as soon as an individual is employed in an organisation and its purpose is to introduce the newly recruited employee.

Oriented tillage: Tillage operations which are oriented in specific paths or directions with respect to the sum, prevailing winds, previous tillage actions or field base lines.

Orographic cloud: Cloud, which is formed by, forced uplift of air over high ground. The reduction of pressure within the rising air mass produces adiabatic cooling and, if the air is sufficiently moist, condensation.

Orographic rain: Rain, which is caused, entirely or in major part, by the forced uplift of moist over high ground. The formation of orographic cloud is followed, in the event of continued uplift of the air by precipitation.

Osmosis: The diffusion of water across a differentially preamble membrane from a region of greater water concentration to a region of lesser water concentration.

Osmotic effect: The force a plant must exert to extract water from the soil. The presence of salt in the soil water increases the forces the plant must exert.

Osmotic pressure: The pressure developed due to unequal concentrations of salt separated by a semi-permeable membrane (or plant cell wall). Water will move from the side of lower salt concentration or higher free energy through the membrane to the side of high salt concentration or low energy. This water exerts additional pressure termed as osmotic pressure termed as osmotic pressure (numerically equal and opposite to solute potential).

Out crop: The actual edge of the inclined stratum at the surface ground.

Out cross: The mating of hybrid with a third parent, also an off type plant resulting from pollen of a different sort contaminating a seed field.

Outer crop/guard crop/borer crop: The crops which are grown around the field boundaries in narrow strips with objectives of protecting the main crop from stay cattle and producing livestock feed and/or seed are called out or guard crops, e.g., *Sesbania* or *Leucaena* on boundaries of field/plantation crops and caster around spring planted sugarcane.

Outlets: These are a feature of bunding systems and serve the purpose of a drainage channel to carry excess water away. Outlets are fed by the numerous drains that enter them. Outlets are of many types: Vegetated waterways, Waste weirs (escape weirs), Pipe outlets. Waste weirs are expensive and are suitable only for small areas. Outlets for gully control are weirs, drop spillways, chutes and flumes.

Output: Refers here to farm outputs, *i.e.*, products of functions which are obtained by means of farming activities and which are composed by the farm household, re-invested in farming (used as internal inputs) or externalised (exchange or sold).

Ovary: The gonad of a female animal; in plants, the basal part of pistil containing ovules.

Oven dry soil: Soil is considered to be oven dry when it has reached equilibrium with the vapour pressure from an oven at 105°C. The oven dry condition corresponds to a relative humidity of approximately 0 per cent of *p*f near 7.

Over irrigation: The application of more water than is necessary for the needs of vegetable, resulting in less of water through seepage and leaching.

Over seeding: In clouds seeding, the hypothetical artificial production of an excessive number of the ice crystals.

Overseas training: Training in a foreign country to acquire expertise in areas in which expertise in not easily available locally.

Ovule: In plants; the structures present in ovaries which produce megaspores and embryo sacs and later develop into seeds. The body, which after fertilization becomes seed.

Oxylophytes: Plants tolerant to high acidic soil conditions.

Ozone: The tritanic form (O_3) of oxygen, of molecular weight 47.998, which is present in the atmosphere in very small amounts ranging from about 0.2 to 0.5 cm equivalent thickness at normal temperature and pressure.

P

P: The symbol denotes the parents of an individual or the parental generation.

p.s.i.: The pressure is usually measured as pounds per square inch (p.s.i.).

Pachytene: A substage of prophase I of meiosis following zygotene and preceding diplotene; often *n* chromosomes are visible as long closely paired structures (bivalents).

Paira crop (uthera): It is a crop sown broadcast in the standing crop of low land rice before its harvest where the residual moisture is used for the establishment of uthera crop, *e.g.*, lathyrus, gram, lentil etc. in standing crop of rice.

Palindrome: A DNA segment in which the base-pair sequence reads the same in both the directions from a point of symmetry within the segment.

Palmate: Compound leaves in which leaflets arise at the same point, *i.e.*, at the apex of the petiole.

Panel discussion: A panel discussion is an effective method for discussion of any particular topic by a panel of three to five experts representing, different areas of the same subject.

Panevaporation: It is the value of evaporation from an open pan evaporimeter, usually from standard USWB, class 'A' pan evaporimeter, during a day or specified period to a place.

Panicle: An inflorescence in which the branches of the primary axis are recemose and flowers are pedicellate, as in tree.

Panmictic Population: A random mating population.

Panmixis: *See, Random Mating.*

Papain: The latex exudated from the unripe papaya fruit having proteolytic activity.

Papilionaceous: Belonging to the sub-family papilionaceae.

Paracentric inversion: An inversion in which the centromere is not included in the inverted segment.

Parameter: A characteristic of a statistical population, such as its average, its standard deviation or variance and are often cannot be measured directly but are estimated from sample of the population.

Parasite: An organism that lives in or on another organism (the host), from which it obtains its food.

Parboiling (rice): Parboiling treatment before milling and it involves soaking the Paddy in water, steaming it in hot water and drying it in the sun. Parboiling makes husk removal easier, toughens the grain which results in reduced breakage during milling and polishing and improve the keeping quality of rice. Also parboiled rice retains more nutrients while processing and cooking.

Parietal placentation: Placenta bearing ovules develop on the inner wall of the one-chambered ovary corresponding to the confluent margins of carpels.

Parthenocarpy: The development of fruit without fertilisation phenomenon of seedless fruit production.

Parthenogenesis: The development of a new individual from a germ cell without fertilization.

Participatory rural appraisal: PRA is a method of extension which enables outsiders to learn about any community and to make use of indigenous technical knowledge (ITK) to supplement their analysis. It is a facilitate technique, in which knowledge is obtained and articulated by investigation, presentation and analysis carried out by rural people themselves. PRA allows them to own the information and identify their priorities themselves. PRA is a departure from traditional methods of data-gathering and planning which are top-down in approach. This method uses locally available materials as tools for interaction, to make the villagers feel at ease and thereby increase participation by transferring initiative to rural people. PRA generates a rapport between the local people and the development authorities. It elicits, presents and cross checks information quickly and brings rural people face to face with planners, generating a dialogue in which local needs are taken into consideration without enforcing outsider opinion. This decentralisation enables local people to take command of their resources and determine what fits their needs.

Participatory technology development (PTD): The process of combing the indigenous knowledge and research capacities of the local farming communities with that of research and development institutions in an interactive way, in order to identify, generate, test and apply new techniques and practices and to strengthen the existing experimental and technology management capacities of the farmers. Also referred to as people-centred Technology Development.

Parts per million (ppm): Design the quantity of a substance contained in a million parts of a mixture of solution in carrier, such as air or water.

Pastoralism: The rearing of livestock which graze primarily natural pasture. Nomadic pastoralism refers to a more or less constantly wandering mode of livestock keeping. In transhumsant pastoralism the herds are moved seasonally or periodically between two regions of differing climate regimes (*e.g.*, mountain/valley); the pastoralist occupy a permanent residence in at least one of this region. In sedentary Pastoralism the animals are kept year-round near a permanent residence.

Pasture: The term pasture is applied to a grazed plant community usually of several species often of diverse botanical type.

Pasture land: It is an area of land covered with grasses or other herbaceous forage plants for use by grazing animals. If properly developed, pastures offer one of the most effective and economical means of conserving and enriching the soil. They provide soil cover and surface mulch, preventing further erosion. Lands of classes II, III, IV, V and VI can be converted into pastures. Pasturelands can be protected by social fencing and improved through the introduction of improved grass varieties and periodic fertilization. Marginal lands which cannot sustain agricultural crops particularly during drought years, can be developed into dependable pastures through the use of contour trenches and contour furrows. Controlled grazing also helps build forage resources native pastures, which lack legume component, can be artificially renovated by sowing leguminous species to provide for more nutritional efficiency. Research indicates that application of 20-25 kg nitrogen increases dry matter yield of grasses considerable. Similarly 30-40 kg of P_2O_5 gives good response of the legume component. Among the grasses for rainfed areas, Dictianthum, Urochloa, Lasiurus, Sehima and Cenchurus have been found to be ideal. Different legumes like Dolichos, Leucaena, Macroptilium, Desmodium, Clitoria, Tephrosia, Indigofera. Crotalaria, Cassia and Stylosanthes have been evaluated with or without grasses for their potential yields. It is almost excellent in terms of persistence, nutritive value and palatability.

Pathogen: Any micro-organism or virus that lives and feeds (parasitically) on or in a larger host organisms and thereby injures it.

Pea berry: A coffee berry with single seed instead of normal two seeds.

Pedagogy: The science of child learning is called pedagogy.

Pedicel: The stalk bearing flower of an inflorescence.

Pedigree: A record of the ancestry of an individual, family or strain.

Pedology: The science dealing with the laws of origin formation and geographical distribution of the soil as a body in the nature.

Peduncle: Stalk of a flower cluster or inflorescence.

Penetrance: The ability of a gene to express itself in all the individuals which carry it in the appropriate genotype (complete penetrance); many genes fail to do so (incomplete penetrance).

Peptide bond: The covalent chemical bond formed due to the reaction of the amino (—NH_2) group of one amino acid with the carboxyl (—COOH) group of another amino acid; it links the various amino acids of a polypeptide with each other.

Per cent concentration: It is similar to ppm except that it is expressed as a percentage, *e.g.*, 1,000 ppm is equivalent to 0.1 per cent and 5,000 to 0.5 per cent.

Per cent germination: The proportion of the number of seeds, which have produced seedlings, classified as normal under the conditions and period specified.

Percolation: It is the downward movement of water through the interstics of rock of soil under gravity or hydrodynamic pressure or both under saturated or nearly saturated condition.

Percolation tank: It is a runoff storage reservoir used for recharging ground water supply and increasing water levels in nearby wells. The location of a percolation tank must be such that the water from surrounding fields will fill it. The water is expected to infiltrate and recharge the downstream area.

Perennial: A plant that lives for three or more years which normally flowers and fruits at least in its second and subsequent years.

Perfect flower: A flower having both male and female parts.

Perfect gas: A hypothetical gas, which obeys the gas laws of Boyle and Charles perfectly. For practical purposes the gases which compose unsaturated air may be considered perfect gases.

Performance: The performance indicates the behaviour that is expected as evidence that the training have achieved the intended outcome.

Performance analysis: The approach that deals with determination of the performance context of training.

Perianth: A collective term for calyx and corolla.

Pericarp: The outermost covering of a fruit.

Pericentric inversion: An inversion in which centromere is located within the inverted segment.

Peri-clinal chimera: A type of chimera in which a tissue from one genotype occurs as a relatively thin layer and completely cover the genetically different core.

Periderm: A part of the bark (of woody plants) that contains tightly packed suberized cells (with no intercellular spaces) which are composed of fatty acids, lignin, cellulose and terpenes; it is composed of the phellogen, the phellem (cork) and the phelloderm.

Perigynous: The condition where the thalamus grows upward the ovary in the form of a cup, carrying on its rim the sepals, petals, and stamens (petals and stamens are attached to the margin of the hypanthium).

Permaculture: A consciously designed, integrated system of perennial or self-perpetuating species of crops, animals or structures (IASA 1990).

Permanent cropping system: An agricultural system in which crop plants are grown continually on the same plots of land, either several crops in rotation or a single crop for several years; disturbance of microclimate and of water relations is permanent, nutrients and organic matter are replaced by inputs from outside.

Permanent pasture: Pasture of perennial or self-seeding annual plants maintained for several years for grazing.

Permanent wilting point: The moisture content in percentage of a soil at which nearly all plants wilt and do not recover in a humid dark chamber unless water is added from an outside source. This is the lower limit of available moisture range for plant growth. Below the wilting point extraction of moisture continues for some time but growth ceases completely. The force with which moisture is held by the soil at this point corresponds to 15 atmospheres.

Permeability: It is the property of the soil to absorb and retain water. It is also the readiness with which air, water or plant roots penetrate through the horizon of the soil. It is a broad term and can be further specified as hydraulic conductivity and intrinsic permeability. 'K' in Darcy's law represents the rate of flow of water in a porous medium under a quality and quantity of water but the value of intrinsic permeability, the property of porous medium remains constant for any fluid.

Permeameter: A device for measuring the permeability of soil or other materials.

Persistence: In meteorology, a term used of a synoptic feature or meteorological condition that is unusually long-lasting.

Persistence forecast: A type of forecast often used as a basis of comparison in the assessment of the success attained in forecasts made by conventional methods. In which the assumption is made that meteorological conditions during the forecast period will remain unchanged from those obtained at the beginning of the forecast period.

Pest management: Manipulation of pest or potential-pest

populations so as to diminish their injury or render them harmless (IASA 1990).

Pesticides: They comprise a large number of various chemicals substances used to control arthropod pests and infectious diseases, improve the fertility of soils and prevent the growth of weeds. They include insecticides, herbicides, nematicides and fungicides.

Petrophyte: A plant able to grow on rocks.

***p*F:** It is the logrithm of height in cm of a water column that represents total stress with which water is held by a soil.

***p*H:** It is a measure of the acidity or alkalinity of soil. It is expressed in terms of a *p*H scale from 0-14. A *p*H of 7 indicates neutrality, higher values indicate alkalinity, and lower values acidity. On the basis of *p*H measurements, the degree of soil acidity is approximately indicated as:

pH Values	*Reaction*
Above 8.5	Strongly alkaline
7.5 to 8.5	Alkaline
6.6 to 7.5	Nearly neutral
6.1 to 6.5	Slightly acidic
5.6 to 6.0	Medium acidic
5.1 to 5.5	Strongly acidic
4.5 to 6.0	Very strongly acidic
4.5 and below	Extremely acidic

***p*H soil:** the negative logrithm of the hydrogen-ion activity of a soil. The degree of acidity or alkalinity of a soil as determined by means of a glass quinhydrone or other suitable electrode or indicator at a specified moisture content or soil to water ratio and expressed in terms of the *p*H scale, from 0 to 14.

Phage: *See, Bacteriophage.*

Phased planting/staggered planting: It is planting/sowing of a crop spread over an optimum period of planting either to minimize risks or to use labour/machinery more effectively or to minimise competition (in intercropping) or to prolong the period of supply to the market or the factories.

Phenocopy: An individual whose phenotype has been modified by the environment to such an extent that it is similar to that produced by another, usually a mutant, allele.

Phenology: The study of the timing biological events, the causes of their timing with regard to biotic and abiotic factors and the inter-relation among the phases of the same or different species.

Phenotype: Physcial expression or outward appearance of a genotype, which is subject to change with the environment.

Phloem: The principal food-conducting living tissue of the vascular plant, basically composed of sieve tube, companion cells, fibres and sclereids; transports metabolic compounds from the site of synthesis or storage to the site of utilization.

Phosphate mineralization: Solubilization of insoluble organic phosphatic compounds (*e.g.*, phytin, hecithin etc.) by microorganisms by virtue of their organic acid production.

Phosphates: Various salts of phosphoric acid applied to the soil as phosphatic fertiliser to supply phosphorus to crop.

Phosphorus deficiency: Deficient plants usually have dark green more erect leaves than normal. In some crops, leaves show orange or purplish discolouration. Phosphorus deficiency can occur in strongly acid, acid sulphate, peat and alkali soils, and is more available in flooded soils than in dryland soils.

Photochemical reaction: A chemical reaction, which proceeds at the expense of, absorbed light, for example photosynthesis.

Photoperiod: The relative length of time that a plant is exposed to light especially as it affects the life cycle and physiological process of the plant. The length of daylight necessary for flowering.

Photoperiodism: The effect of day length on flowering, dormancy, vegetative development, leaf fall and other aspects of plant behaviours.

Photophosphorylation: The process of ATP formation, in plants, in the presence of light.

Photorespiration: Light-dependent oxygen uptake and carbon dioxide production (distinct from mitochondria respiration, and is primarily more in C-3 plants and is very rare in C-4 plants.

Photo-spectrometer (emission spectrometer): An instrument that disperses the light from a source and produces a spectrum indicating relative intensity as a function of wavelength.

Photosynthesis: Conversion of light energy into chemical energy by plants with the help of chlorophyll, carbon dioxide and water.

Photosynthetic efficiency: The percentage of total solar energy assimilated by plants. The energy fixed by plants expressed as percentage of the incoming solar energy.

Phylogeny: The evolutionary history of a species.

Physical dormancy: Dormancy due to impermeable seed coat.

Physiological antagonism: The type of antagonism that occurs when two components of a mixture, acting at different sites, counteract each other by producing opposite effects on the same physiological process.

Physiological disorder: Disorders are due to abnormal environmental conditions, improper nutrition and without involvement of primary parasite.

Physiological maturity: Attainment of final stage of biological function by a plant part or the plant as a whole.

Phytoplankton: Microscopic floating plants.

Phytoxic: Injurious to plant life or life processes.

Pink berry of grapes: A disorder of grape wherein some berries develops off pink colour which turns black within a day after harvesting and incidence increases with rising temperature.

Pinnate: Compound leaf with the leaflets arising along an axis – like a feather.

Pisciculture: It is the breeding, hatching and rearing of fish in water harvesting structures, under controlled conditions. Pisciculture can be adopted as an activity to supplement the food requirement and income of farmers in rainfed area. The introduction of pisciculture should however be restricted to areas where water harvesting is possible and areas where water is available for at least 10 months a year. The activity is ideal in heavy and medium rainfall areas with impervious soils. Low rainfall areas and porous soils are not suitable for pisciculture. Pisciculture is of two types: Capture fisheries and culture fisheries.

Pistil: A unit of the gynoecium comprising stigma, style and ovary.

Pistillate flower: flower with pistils but no stamens.

Plagiotropic: The term used primarily for roots, stems or branches to describe growth at an oblique or horizontal angle.

Plant disease: A plant disease is a physiological or structural deviation from the normal that is detrimental to the health of the plant and reduces the economic value. A disease expresses itself in a series of definite symptoms. It could be localised (affecting a single area or structure) or systemic (where the infection spreads through the entire plant body), both of which cause considerable disturbance to plant metabolism. Diseases are aggravated by environmental factors such as temperature, water availability, and soil conditions along with physiological factors such as lack of nutrients and the presence of toxins.

Plant diseases are classified into the Fungal: Viral diseases are most lethal to plants as they spread often through direct contact, seeds, grafts, cuttings and insects. The damage is caused internally to tissues making it difficult to inactivate them.

Plant ecology: The branch of ecology, which deals with the study of plants in relation with the environment.

Plant food ratio: The ratio of the number of fertiliser units in a given mass of fertiliser expressed in the order NPK.

Plant growth regulators: Organic compounds occurring naturally in plants as well as synthetic other than nutrients which in small amounts promote, inhibit or modify any physiological process in plants.

Plant nutrients: The elements taken in by the plant which are essential to its growth and used by its elaboration of its food and tissues. Includes nutrients obtained from fertiliser ingredients.

Plant pests: Insects and nematodes which are harmful to plants are called plant pests. In rainfed areas the farmers have very little access the use of pesticides. This factor, coupled, with the lack of resources, leads to major losses through afflictions every year.

Plant quarantine: It may be defined as a restriction imposed by appropriate authority on the production, movement or existence of plants and plant material.

Plant stand: It is the number of plants per unit area.

Establishment of optimum plant stand is a prerequisite for successful crop production, especially in rainfed area where seed germination is limited, by inadequate moisture and resultant hard crust formation. If the stand is too sparse, soil moisture will be rapidly lost, evaporation from the soil surface and available moisture cannot be used economically. If there are more plants than optimum, then there may be early mortality or under-nourishment. Punjab (Including Delhi and Haryana) 630 mm; Rajasthan east 700 mm; Saurashtra and Kutch 480 mm; Rajasthan west 310 mm; Rayalaseema, interior Karnataka 560-700 mm; (including large pockets of Madhya Pradesh, West Tamil Nadu and Marathwada).

Planting ratio (seed production): It is recommended ratio in which male and female parental lines are planted to make a crossing block in hybrid seed production.

Plaque: The clear area or zone produced by a virus on an opaque lawn of the host bacterium; generally, plated on an agar medium in a petri plate.

Plar compounds: Characterised by electrical charges.

Plasma gene: A gene located in a cytoplasmic organelle, *e.g.*, plastid, mitochondria; responsible for cytoplasmic inheritance.

Plasmid: In bacteria; a DNA molecule other than the bacterial chromosome that is capable of independent replication and transmission; it may exist independent of or become integrated into the bacterial chromosome; generally, not essential except under specific environments.

Plastid: In green plants and some protozoa; an organelle surrounded by two concentric membranes, has DNA, and usually differentiates into typical chloroplasts when exposed to light.

Pleasant voice: A voice, which can convey warmth and listeners, would not be tried of listening for a longer duration of time.

Pleiotropic: A gene controlling more than one character.

Pleiotropy: A single gene affecting more than one character. (*Adj.*, pleiotropic).

Plot experimental: An experimental unit to which a treatment is applied (It may be single, small area of soil containing several plants, a strip through a field or a whole field).

Plough sole: A compacted layer at the bottom of the zone of ploughing.

Plough sole placement: A method of fertiliser application in which fertiliser is applied in a continuous band in the bottom of the furrow, in the process of ploughing each band is covered as the next furrow is turned.

Plumule: The primary bud of a germinating seed which forms the stem.

Pluvial rice lands: Lands, which are well drained without free ground water within the rooting zone of the rice plant and are generally unbounded.

Polar Mutation: A mutation which affects the expression of other genes located downstream the mutant gene but in the same operon of prokaryotes; due to the formation of a nonsense codon.

Polarity: The electrical phenomenon of a molecule or an ion, on the basis of which the chemicals can be divided into polar and non-polar compounds.

Pollarding: Tree whose stem has been cut off in order to obtain a flush of shoots, usually above the height to which the browsing animals can reach.

Pollination: The transfer of pollen from anther to stigma.

Pollinator: The agent of pollination.

Pollinizer: The producer of pollen; the variety used as a source of pollen for cross pollination.

Polyculture: Growing of several species of plants in a given unit area.

Polyembryony: The production of more than one embryo per ovule.

Polygamodioecious: The condition in which plants are functionally dioecious but a few flowers of opposite sex or a few bisexual flowers appear on all parts during bloom period.

Polygamomoniecious: The condition in which plants are basically Monoecious but a few flowers of the opposite sex or a few bisexual flowers appear on all parts during the bloom period.

Polygamous: Bearing unisexual and the bisexual flowers on the same plant.

Polygenes: Genes controlling a quantitative character having small, similar and additive effects.

Polymorphic: Having more than one form in the same population.

Polymorphism: Two or (usually) more phenotypes maintained in a random mating population; due to the effects of environment.

Polynucleotide: A macromolecule in which several nucleotides are joined by the phosphodieter linkage, *e.g.*, DNA and RNA.

Polypeptide: A macromolecule in which several amino acids are joined by the peptide bond.

Polyploid: A cell or an organism having more than two genomes; they may be copies of a single genome (*autopolyploid*) or of different genomes (*allopolyploid*).

Polytene chromosome: A giant chromosome made up of several (upto 1,024 or more) copies of itself produced by endomitosis; usually, the homologous are paired as if during pachytene; in *Drosophila*, characteristic bonds present. (*Syn.*, giant chromosome, salivary gland chromosome).

Pome fruit: The fruit type derived from the fusion of the ovaries, calyx cup and floral tube, produced by the apple, pear, quince and other members of the sub-family pomoidaea.

Population: In *population genetics*; a group of interbreeding individuals sharing the same gene pool. In *statistics*; all the possible individuals or observations of one kind from which samples are drawn for various studies.

Population genetics: The branch of genetics concerned with the study of frequencies of genes and genotypes in random mating population.

Population, effective: The number of sexually reproducing individuals in a population.

Position effect: Alternation in the phenotypic expression of a gene due to a change in its position in the chromosome; usually, when a gene is placed within or close to a heterochromatin segment; due to heterochromatinization the gene is not expressed.

Post-emergence application: It is the term employed when weed killers are applied to a crop after it has emerged through the soil surface.

Potassium absorption ratio (PAR): A ratio for soil extracts and irrigation waters used to express the relative activity of potassium ions in exchange reaction with soil. Where the ionic concentrations are expressed in milliequivalents per litre or millimoles per litre.

Potassium deficiency: Symptoms of mid-deficiency are dark green leaves, low tillering and stunting. Severe deficiency results in yellowish orange to yellowish-brown discolouration starting at the tip of older leaf blades and gradually moving towards the base. Necrotic spots may appear on the leaf blade, grain size and weight may be reduced. Potassium deficiency usually occurs in course, sandy, or peat soils and in soils rich in potassium fixing clays.

Potassium fixation: The process of converting exchangeable or water soluble potassium to a form not easily exchanged from the adsorption complex with a cation of a neutral salt solution.

Potassium sulphate: A potassium fertiliser containing 48 per cent K_2O and also supplying 17-20 per cent of sulphur.

Potential crop yield: Yield of a crop obtained at research stations under most ideal environment.

Potogyny: A condition in which stigma is receptive before the pollen of the same flower is shed.

Ppm (parts per million): It refers to the number of parts by weight or volume of a constituent in 1,000,000 parts of the final mixture by weight or volume.

Precipitation: Total amount of precipitation (rain, drizzle, snow hail, fog, condensation, frost) expressed in depth of water which would cover a horizontal plane if there is no runoff, infiltration or evapotranspiration. Unit mm/day.

Pre-emergence application: It is the term employed when weed killers are applied to a crop after it has been sown or planted but before it emerges the soil surface.

Pre-emergence treatment (herbicides): Herbicides applied after seeding but prior to crop or weed emergence.

Preplant incorporation: Any treatment applied and incorporated (blended) into the soil before seeding or transplanting a crop or weed; incorporation usually by tillage.

Pre-planting treatment (herbicides): Treatment applied after the soil has been prepared but before seeding/planting.

Prepotency: The capacity of a parent to impress characters on its offsprings.

Prey crop (smother crop): The crop which is grown with the purpose of eliminating any undesirable plant through physical or its allopathic effect.

Principle of limiting factors (Liebig's law of minimum): It states that when a process is conditioned as to its rapidity by a number of separate factors, the rate of the process is limited by the factor which is at the minimum.

Probability: The chance or likelihood of the occurrence of an event.

Process evaluation: The process evaluation is also known as formative evaluation. This is done to solve the problems that arise in implementation of a programme.

Process skills: Process skills are related to setting goals, developing plans, reviewing and evaluating the process of identifying training need.

Processing: The third stage of experiential learning cycle, which is concerned, with analysis of experiences shared by the trainees.

Productivity: The relationship between the quantity of goods or services produced and the factors used to produce it; farm productivity can be expressed as output per unit of land, capital, labour, time, energy, water, nutrients etc.

Productivity (plant): The capacity of a plant to produce yield or biomass per unit area per unit time.

Progeny: Offspring.

Progeny test: The practice of ascertaining the genotype of an individual by mating it to a individual of known genotype and examining the progeny.

Projection: The process in microlab in which the participants are expected to examine their thinking or imagination through projection.

Prokaryote: Organisms without true nuclei, *i.e.*, the chromosomes are not enclosed by a typical nuclear envelope; they do not undergo meiosis, *e.g.*, viruses, bacteria, cyanobacteria.

Promoter gene: The gene or DNA segment of an operon to which RNA polymerase binds before it initiates the transcription of an operon.

Propagation: The act of process of breeding plants and animals.

Propaqule: Any part of a plant capable of growing into a new plant – a seed, spore, rhizome, leaf, twig or even root part.

Prophase: The first stage of cell division after interphase and before metaphase; the chromosomes become shorter and thicker, and are visibly double-stranded toward and end (mitosis, prophase II of meiosis); chromosomes pair to form typical bivalents (prophase I of meiosis).

Protandry: A condition in which the pollen grains are shed before the stigma of the same flower becomes receptive.

Protein efficiency ratio (PER): A term used to measure the humidity of the atmosphere. It comprises of 2 identical thermometers the bulb of one is dry and the bulb of other is covered with thin film of distilled water.

Protogyny: The stigmas become receptive before pollen is shed.

Proto-oncogene: A normal cellular gene which may change into an oncogene, most likely due to mutation.

Prototroph: A microorganism which is able to grow normally on a minimal medium containing inorganic salts and a carbon source.

Pruning: Partial or complete removal of vegetative or fruiting wood from plant to control its size, remove broken or damaged tissue, after plant shape, remove unnecessary growth or balance growth.

Pseudoalleles: Two alleles each of two or more tightly-linked genes affecting the same function as a result of which they appear as multiple alleles, once an attractive idea for explaining many cases of multiple alleles, *e.g.,* Rh of blood groups; *only a hypothesis, and not widely accepted.*

Pseudogene: An inactive gene similar to a normal gene and occurs as a normal component of an eukaryotic genome; may arise due to mutation following duplication of the concerned gene.

Pseudohermaphrodite: Functional unisexualism in plants which bear perfect flowers.

Psychomotor learning: Learning that deals with learning about physically doing something.

Psychroxerophte: A drought resistant plant growing in cold territories.

Pubescent: Hairy.

Publishing: The second stage of experiential learning cycle where the data generated through the first stage, i.e. experiencing is made available to the entire group.

Puddle: A compact mass of earth, soil, clay or mixture of two or more of them which has been compacted through the addition of water and rolling and trampling and made more or less impervious.

Puddled soil: Dense massive soil artificially compacted when wet and having no results from the tillage of a clayey soil when it is wet.

Puddling: Making a soil impermeable by manipulating and compacting it in standing water, which reduced its apparent specific volume thus facilitates transplanting.

Pulse seed: The edible seed of legumes.

Pure line: The progeny of a single, self-fertilised homozygous individual.

Q

Quadrat: Limited sample area, from which total counts are made, is usually known as quadrat.

Qualitative character: A character showing few distinct classes, governed by one or few oligogenes, and relatively little affected by the environment.

Quadruple cropping: Growing four crops a year in sequence.

Quantitative character: A character showing continuous variation, governed by polygenes, and markedly affected by the environment.

Quantitative inheritance: The inheritance of quantitative traits due to polygenes.

Quantum yield: Those value which separate a data set into four equal parts.

Quasi-bennial oscillation: A well defined oscillation of the zonal wind component in the equatorial stratosphere, the 'period' being about 27 months. It is also termed 'stratosphere oscillation'. The range of the fluctuations, greatest at a level of about 25 km, decreases with distance from the equator. There is phase lag from higher to lower levels. At a given level the phase does not appear to vary with longitude.

Questioning: The technique of asking questions and securing the involvement of participants.

Quiescence: The condition in which the seed can germinate immediately upon the absorption of water in the absence of any internal germintion barriers.

Qyadric response: The output goes on increasing upto a certain level of input and then starts decreasing when input is increased further. In such cases the response is said to be quadratic and is given by: $Y = a + bx + cx^2$.

R

R_1 R_2: The first and second generations after irradiation.

Rabbing: The practice of burning the paddy nursery by spreading dried cowdung, twigs and dry grass which is common in the rice tracts of western India which has many advantages other than weed control.

Rabi crops: Crops, which are grown during the winter season (October-November to March-April) with irrigation or on conserved, soil moisture, *e.g.*, wheat, barley, oats, chickpea, safflower etc.

Rabi season: Season during which winter crops like wheat, barley, mustard, gram, linseed, etc. are cultivated and usually it extends from October to April.

Raceme: A simple elongated indeterminate inflorescence with pedicillate flowers.

Rachis: The main axis of an inflorescence or the axis of a pinnately compound leaf to which leaflets are attached.

Radar: A system of detection and location of 'targets' which are capable of reflecting high-frequency radio waves (microwaves), generally in the wavelength range from a fraction of a centimetre to some tens of centimetres.

Radial: A young root.

Radiation balance: The resultant flux of the solar and terrestrial radiation through a horizontal surface. It is considered positive in the flux downward exceeds that upwards. It is also termed 'net radiation'.

Radiation fog: A common type of fog which forms overland on night characterised by light, wind, clear sky, and moist air in the lower levels of the atmosphere.

Radicle: The lower part of the embryo which forms the main root.

Radiosonde: A small radio transmitter by means of which observations, usually of pressure, temperature and humidity, may be obtained from the upper atmosphere.

Rain gauges: A rain gauge is an instrument for measuring rainfall or other forms or precipitation. Gauges are located at predetermined points in an area and the average rainfall value determined. The precipitation measured by a rain gauge is that which is supposed to be distributed over a horizontal impervious surface and is not subject to evaporation. The amount of precipitation is expressed in terms of depth of either over an area in inches or millimetres and the rate or intensity of rainfall is expressed as inches or millimetres per hour. There are three types of rain gauges: Non-recording or standard gauges, Recording rain gauges, Storage gauges.

Rainfall: The total liquid product of precipitation or condensation from the atmosphere as received and measured in a rain gauge in past 24 hours at 03 UTC. Classification of rainfall is as follows: Very heavy–rainfall more than 12.5 cm; Heavy-rainfall 6.5 to 12.5 cm; Rather heavy–rainfall 3.5 to 6.5 cm; Moderate–rainfall 0.8 to 3.5 cm; Light rain–rainfall 0.3 to 0.7 cm; Very light rain–rainfall up to 0.2 cm; No rain–when there is zero rainfall.

Rainfall distribution: Spatial distribution of rainfall is highly variable. The following criteria are used in forecasting practices. *Isolated:* When rainfall occurs 25 per cent or less of the total area under observation. *Scattered:* When rainfall occurs 26 to 50 per cent of the total area under observation. *Fairly widespread*: When rainfall occurred 51 to 75 per cent of the total area under observation; *Widespread*: when rainfall occurred more than 75 per cent of the total area under observation.

Rainfall zones: It is a graph that indicates the amount of rainfall received in an area over a specific period. It could be yearly rainfall for the past 50-100 years to forecast runoff flow to design a reservoir, or daily, weekly and monthly rainfall providing an idea about the quantity, duration and spread of rainfall. Development plans may be drawn using information from rainfall distribution graphs developed at research and meteorological centres, providing information about rainfall period, duration and the intensity of each period. Rainfall distribution graphs help identify proper time of monsoon. The annual rainfall in terms of the standard meteorological week is taken for studying the rainfall distribution. An area with uniform rainfall distribution constitutes a rainfall zone. These are important for the purpose of planning crops. Information on the sequence of the wet and dry weeks can help to formulate a strategy to be adopted for a given area for successfully implementing different crop programmes on a sustained basis. Low rainfall zones receiving less than 600 mm per annum, medium rainfall zones receiving between 600 to 750 mm of rainfall and high rainfall zones receiving more than 750 mm of rainfall. Nearly one-third of the total area of the country receives a rainfall of less than 750 mm.

Rainfed farming: Growing of field crops in areas where crop growth is dependent entirely on the rains is called rainfed farming. Rainfed systems include about 88.7 million hectares of arable land in our country today. This is besides about 20 million hectares of barren and uncultivable land. 15 million hectares of cultivable waterland and 12 million hectares of permanent pastures and grazing land. Degradation of such vast rainfed lands is a major cause of ecological imbalance in India. Integrated development of such lands, holds the key to the ecological health of our country.

Rainfed farming: Growing of field crops with rain water received during the crop season (rainfall usually > 800m) under humid and sub-humid climate and the crops may face little or no moistures stress during their life cycle.

Rain-gauge: Instrument for measuring the depth of water from precipitation supposed to be distributed over a horizontal impervious surface and not subject to evaporation.

Ranching: Grazing of cattle extensively on a fully commercial basis, controlled both by herding and fencing of territories, environmental pressure adequate ranges quality.

Random genetic drifts: the fluctuations in gene frequencies in small random mating populations due to sampling errors.

Random mating: A system of mating in which a kind of male gamete has the same opportunity of uniting with any kind if female gamete of the population; or an individual of a genotype in the population has equal probability of mating with an individual of any genotype present in the population.

Random mating population: A population in which random mating is the normal mode of reproduction.

Randomization: An act of assigning the treatments to different plots by a random process.

Range management: Management of range to produce maximum forage for sustained use without jeopardising the other resources or uses of land.

Rational grazing: Dividing a large pasture into several small Paddocks and allowing the cattle to graze each, periodically perhaps every four weeks.

Ratoon: The emergence of new tillers from the root/crown of a plant or regeneration of stalks from the previous year crops, as is done in brinjal, chilli, tomato, sugarcane, sorghum etc.

Ratoon cropping: The cultivation of an additional crop from the re-growth of stubble of previous main crop after its harvest, thereby avoiding reseeding. Replanting such as in sugarcane, sorghum, rice, fodder grasses etc.

Raw crop: Crops, which are sown in rows, keeping uniform spacing throughout the field *e.g.*, cotton.

Recession phase (irrigation): Portion of total irrigation time between the beginning of recession at the upper end and the disappearance of the water from the field surface time elapsed during this phase is known as recession time.

Recessive: The member of an allelic pair which is not expressed when other member (dominant) occupies the homologous chromosomes, marked by its allele.

Reciprocal crosses: The two crosses in which the same two parents are involved, but the strain which serves as the male parent in on cross in used as the female parent in the other, and *vice versa.*

Reclamation: It refers to the process of making land more suitable for intensive use by changing its general character through operations like Drainage, Irrigation for arid and semi-arid lands. Rejuvenating badly eroded land by using vegetative barriers or checks and dams.

Reclamation (land): The operation or process of changing the condition or characteristics of land which cannot be utilised to its full potential otherwise, *viz.*, irrigation of land, drainage of swamp or waterlogged land, gypsum application to alkali soils, liming of acid soils etc.

Reclamation (saline soils): The process of removing excess soluble salts or excess exchangeable sodium from soils.

Recombination: Usually, in the case of linked genes or characters; the production of new combinations of genes or characters not present in the parents of a cross or generation; involves crossing over.

Recon: The unit of recombination; is represented by one nucleotide; the part of gene within which recombination cannot take place.

Recording rain gauges: They are used to collect information on rainfall intensity and volume. These gauges are equipped with recording devices which give a continuous record of the amount of rainfall, the time of start to the time of cessation; the duration and the intensity.

Recurrent parent: The parent to which the F_1 and back cross plants are crossed.

Recurrent selection: Reselection generation after generation, with inter-breeding of selection plants.

Recycling: In the context of agricultural hydrology, recycling is the application of water, collected from a particular drainage area to the same area for agricultural production. The objective of recycling in rainfed areas is to consume as much rain water as possible and use it for direct irrigation or supplemental irrigation. Harvested water is a scarce commodity and should be sparingly used only for life-saving irrigation at the most critical stages of crop growth or for extending the cropping season thus making inter or sequential cropping possible. Nurseries and vegetables farming also can be taken. The approach adopted in harvested rain water recycling is therefore different from the conventional concept of irrigated farming. Where adequate water supply is available for maximising crop production. Runoff varies from 10-40 per cent of the total rainfall. Of this atleast 30 per cent can be harvested into farm ponds and water so stored can be recycled to the donor areas.

Red soils: The colour of the soil is due to the wide diffusion or iron rather than a high proportion of it. The soils grade from poor, thin, gravelly and light coloured to deep soils varying from the plains to the valleys. They are generally poor in nitrogen, potash and humus. They are poorer in lime, phosphorus, iron oxide and phosphorus than regur soils. The clay fraction of red soils is rich in kaolintic mineral.

Reduced tillage: A tillage system in which the primary tillage operation is performed in conjunction with special Planting procedure in order to reduce or eliminate secondary tillage operations.

Reducing and non-reducing sugars: Sugars possessing a free aldehyde or ketone group are reducing sugars, such as maltose. When the above groups are locked upon a glycoside linkage in the formation of complex sugars, the resulting sugars are said to be non-reducing, for example, sucrose.

Reduction division: The first division of meiosis when the two chromosomes forming each bivalent separate and move to the opposite poles.

Reflective observation: The stage of learning cycle where through observation, listening and reflective behaviour.

Refresher training: Training provided to impart specialized subject matter knowledge to the employees so that they remain updated with regard to latest developments in the field.

Regeneration: Sprouting of suitable plant species following cutting and/or grazing. Synonymous to re-growth.

Regression: It is a quantitative measure of relationship between two variable which helps to predict the value of one variable (dependent) from the known value of the other (independent) variable.

Regression coefficient: A numerical measure of the rate of change of the department on the independent variable.

Regression line: Straight lines which represents the relationship between two variables.

Regulator gene: The gene of an operon which codes for the repressor of the operon, *e.g.,* the gene i of *E-coli lac* operon.

Relative crop intensity index: It determines the amounts of area-time alleviated to one crop or group of crops relative to the area-time actually used in production of all crops.

Relative crowding coefficient (*k*): It is a measure of the relative dominance of one component crop over the other in an intercropping or mixed cropping system. The coefficient (*k*) is determined separately for each component crop.

Relative density (RD): The population of a particular species expressed as a percentage of all the species in a habitat thus RD = Density of a particular species/Total density of all the species × 100.

Relative dominance: The actual area occupied by a species expressed as a percentage of total area occupied by all species in a given land area. Total basal area of species 'a'/Total basal area of all + 100 species.

Relative growth (RGR): An index of the amount of growing material per unit dry weight of plant per unit time. W_1 and W_2 are the total dry weight at time t_1 and t_2 respectively; the mean value of RGR for the interval t_1 to t_2 is given as Loge W_2 – loge $W_1/t_2 - t_1$. It is also called efficiency index (y) and can be expressed in g/g/day.

Relative net return (intercropping): It is the ratio between the monetary value of base and intercrop plus differential cost of cultivation (intercropping-monoculture of base crop) and monetary value of base crop in monoculture.

Relative spread index: It represents an area of the crop expressed as percentage of the total cultivated area in the zone divided by the area of the crop expressed as percentage of the total cultivated area in the country, time hundred.

Relative yield index: It represents the mean yield for a crop in a region divided by the mean all India yield of the same crop time hundred.

Relay cropping: Seeding/planting of succeeding crop after flowering and before the harvest of the standing crop, it is analogue to a relay race where on crop hands over the baton (land to the next crop in quick succession, *e.g.*, sowing in standing crop of maize.

Replication: The process of production of new copies of a macromolecule by exactly copying the preexisting macromolecule, *e.g.*, DNA → DNA, RNA → RNA.

Replication (experiment): Repetition of the same experimental material several times to ascertain the estimate of experimental error.

Replication plot: Areas of agriculture field in which farmers communities themselves have adapted or adopted agricultural technologies.

Replicon: A segment of DNA capable of undergoing replication; it has a unique 'origin' site at which the replication begins.

Repressible enzyme: An enzyme that is produced by the cell as long as its end-product does not accumulate to a threshold concentration.

Repulsion phase: A linkage between the dominant allele(s) of one (or more) gene(s) and the recessive allele(s) of another (several other) gene(s); in such a case one parent involved in a cross contributes the dominant allele of one (or more) gene(s), while the second parent provides the dominal allele(s) of the other gene(s).

Research: Studies or inquiry involving critical and exhaustive investigation, including experimentation and design to discover new facts.

Resistance: Ability of living organisms to survive the disruption of life process caused by pesticides, disease, drought etc., which would normally cause the death of together similar organisms (IASA 1990).

Resistance factor: A plasmid which carries genes that confer to its host bacterium resistance to one or more antibiotics.

Resource enhancing: Improving the quality of existing resources, referring here to all raw materials, energy sources and human capabilities which can be used for farming.

Resource Inventory: Resource Inventory is a record or tally of the resources of a watershed. It is based on the information collected through the resource survey (*see* resource survey). An inventory provided information on the different resources of a watershed including, land, water, human, livestock and infrastructural resources.

Resource survey: For preparing an action plan to develop the components of a watershed, it is essential to collect information on the basic resources of the area under a watershed programme. Resource survey help identify the present land use and the potential of land, water, human and livestock resources as well as income generating activities and infrastructural support of a watershed. It also helps to update the information contained in village maps and toposheets, like patterns of ownership. Similarly there could be changes in the course of drainage lines which have to be checked up by physically going around the watershed area. These changes need to be incorporated in the map after physically verifying it on the field.

Respiration: The process by which energy is acquired in a living plant organism, by the breakdown of complex organic molecules, with the release of waste carbon dioxide and water.

Response curves: A close function relation between the input x (nutrients) and output Y (production) can be expressed in some algebraic equation in the form $Y = f(x)$; (*i*) $Y = a + bx$ – Linear response; (*ii*) $Y = a + bx \times cx^2$ – Quadratic; (*iii*) $Y = y + d$ (1-10 – kx) – Mistscherluch equation; (*iv*) $Y = Ax^6$ – Cobbs-Douglas or power function; (v) $Y = ae^{bx}$ – Exponential or logarithmic curve.

Restitution: When a chromosome breaks and the two broken ends thus produced join in their original position to yield the original chromosome.

Restitution nucleus: A nucleus which contains all the

chromosomes present in a cell following the failure of a cell to undergo complete division.

Restore line: An inbred lines that when crossed on a male-sterile strain, causes the resulting hybrid to be a male fertile and produce pollen.

Restriction enzyme: An endonuclease which cleaves a DNA molecule either within or close to a site having a specific base sequence.

Resurgence: The increase in a pest population after it is freed from natural controls, most commonly following the application of a pesticide that destroys its natural enemies (IASA 1990).

Retentivity: It is the ability of the soil to retain water, without allowing it to runoff.

Retraining: Retraining refers to efforts designed to prepare an individual for a new assignment or a broadened aspect of the old speciality.

Retting: Soaking flax, hemp, jute or other fibrous plant material in water to facilitate separating the fibres from the stalk for further processing.

Reunion: The joining of broken chromosomes ends, following chromosome breakage, in ways other than they were originally united.

Reverse mutation: The mutation of a mutant allele back to the wild type allele.

RFLPs (Restriction Fragments Length Polymorphism): A genetic difference among individual that is affected by comparing DNA fragments released by digestion with one or more restriction enzymes.

Rh factor: An antigen present on the RBCs of some human beings which is antigenically similar to an antigen found in *Macaca rhesus* monkeys.

Rhizome: The horizontal, slender, underground root-like stem capable or sending out roots and leafy shoots.

Rhizosphere: The portion of the soil in the immediate vicinity of plant roots in which the abundance and composition of the microbial population are influenced by the presence of roots.

Ribonucleic acid: A polynucleotide having ribose as the pentose sugar and ordinarily uracil in place of thymine; genetic material in many viruses and may undergo replication; generally, single-stranded; is of five types: mRNA, tRNA, rRNA, chromosomal RNA, and genetic RNA; produced by the transcription of DNA.

Ribonucleic acid (RNA): Polymeric constituent of all living cells, consisting of a single strand of alternating phosphate and ribose units with the bases adenine, guanine, cytosine and uracil bonded to the ribose, the structure and base sequence of which are determinants of protein synthesis.

Rice glazing: Polished rice treated with talc (and in some cases with glucose).

Rice milling: Involves the removal of hulls and bran from paddy rice. Milling ranges from hand ponding to modern milling equipment.

Ridge: A ridge (of high pressure), also termed a wedge, is an extension of an anticyclone of high-pressure area sown on a weather chart, corresponding with a ridge running out from the side of a mountain. It is the converse of a trough of a low pressure and is generally associated with fine anticyclonic-type weather.

Ridge: *Ridge of watershed*: It is the highest point in the watershed, from where water flows in different directions. *Ridge lines* serve as divisions between watersheds sub-watersheds, micro-watersheds or mini-watersheds. *Ridge in a field*: It is a raised strip of ploughed ground which is gently sloped to permit drainage. Each ridge acts as an obstruction to the flow of rainwater, which can stand in the furrow or depression, there by allowing more water to infiltrate into the soil. Depending on agronomic requirements, the configuration of ridges, furrows may be 75-150 cm apart, broad ridges, furrows may be 75-150 cm apart.

Ridging: Making long, parallel, raised strips of earth, into which seeds are sown. Ridges are usually made at right angles to the slope. They can serve to increase water retention, reduce soil erosion, bury and compost weeds, and create a favourable environment for the seed by concentrating topsoil and raising the root-growing area of the young plants above the water table.

Ring budding: A method of budding in which a ring of bark is removed from the stock and a similar one containing a healthy bud is removed from the bud stick and then slipped over the rootstock.

Ring neck: A disorder of the avocado which causes drying out and shrivelling of surface areas on the fruit pedicles.

Ringing: Removal of a narrow strip of bark round the periphery of the stem to prevent the downward translocation of food matters beyond that point.

Rogue: An off-type plant in crop field.

Role play: A role play is a technique where one or a group of trainees enacts a real life situation which helps the participants to experience the true feelings of that situation.

Root crop: Crop plants whose surplus or reserve foods are stored primarily in enlarged roots, *e.g.*, sweet potato.

Root nodules: Refers to small swelling on roots of leguminous plants produced as result of infection by nitrogen-fixing bacteria (Rhizobia).

Root pruning: Removal of portions or all of the younger roots with a view to develop extensively branched root systems at the base of the plant, practised in guava to regulate cropping.

Rosette: In many plants profuse leaf development with retarded internode growth to give bunchy or resette appearance.

Rotation: Repeated cultivation of a succession of crops (as sole or mixed crops), possibly in combination with fallow, on the same land. One cycle often takes several years to complete.

Rotation pasture: A pasture used for a few seasons and then ploughed for other crops.

Rough rice (Paddy): Unhulled rice.

Rouging: To remove weeds or off-type or diseased plants from a standing field crop.

Row intercropping: Growing two or more crops simultaneously where one or more crops are planted in rows.

Runner: Trailing stems or branches that take roots and reproduce new plants at the joints or ends.

Runoff: The portion of the precipitation which is not absorbed by the soil but finds its way into the streams after meeting the persisting demands of evapotranspiration interception and other lesses. The portion of precipitation which is not absorbed by the soil and which finds its way as surface flow, after evapotranspiration, interception, infiltration, surface storage, surface detention, chained detention and other losses. Runoff occurs when the rate of precipitation exceeds the rate at which water infiltrates into the soil. The factors affecting runoff are **Rainfall:** Duration, intensity and distribution influence rate and volume of runoff. For instance, intense rain exceeds infiltration capacity by a greater margin than does gentle rain. Thus, total volume of runoff is greater with intense rain, even though total precipitation for both is the same. Intense rain decreases infiltration rate by its destructive action on the structure of the soil surface. **Physiographic factors:** These include size, shape, orientation, topography, geology and cultivation practices in a catchment area. Runoff volume and rate increases as watershed size increases. The cropping pattern in a watershed may determine period at which high runoff may occur. Topographic features such as slope of land, degree of developmental gradient of channels and extent and number of depressed areas affects the rate and volume of runoff. Watersheds with extensive flat areas or depressed areas without surface outlets have lower runoff than areas with steep well-defined drainage patterns. The geologic or soil materials determine to a large degree, the infiltration. Vegetation retards overland flow and increases surface detention to reduce the peak runoff rates.

Run-off water: Water, such as rain, that flows off the land in streams.

R-value: It is the percentage of the crop land actually cropped in a year. Frequencies of cropping are a fallow cycle (R) is equal to crop year time 100. "R" value is synonymous to cropping index.

S

S_1, S_2: The first and second generations after self-pollination (Normally used in cross-pollinated crops).

Saline alkali soil: A soil containing a combination of soluble salts and exchangeable sodium sufficient to interface with the growth of most crop plants. The electrical conductivity and exchangeable sodium of the saturation extract are more than 4 millimhos per centimetre at 25°C and more than 15 per cent respectively. The *p*H is usually 8.5 or less in the saturated soil paste.

Saline soil: A soil containing excess of natural soluble salts to impair soil and crop productivity and has *p*H value of 8.5, EC 4 mmhos per cm at 25°C and ESP 15 per cent.

Salinization: The process of accumulation of soluble salts in a soil usually chlorides and sulphates.

Sand: Small rock or mineral fragments having diameters ranging from 0.05 to 2.0 mm.

Sand storm: An ensemble of particles of dust or sand energetically lifted to great heights by a strong and turbulent and strong wind.

Sandy clay: Soil of this textural class contains 35 per cent or more clay and 45 per cent or more of sand.

Sandy clay loam: Soil of this texture class contains 20-35 per cent clay, less than 28 per cent silt and 45 per cent or more sand.

Sandy loam: Soil of the sandy loam class of texture, which has 50 per cent sand and less than 20 per cent clay.

Sandy soil: A broad term for soil of the sand and loamy sand classes where more than 70 per cent sand and less than 15 per cent clay are present.

Sapling: A young tree, especially a young forest tree trunk with a trunk from one to four inches in diameter; the term is used for young trees that are taken to fields for planting.

Saturated soil: A saturated soil, is one in which all the soil pores including those which in a healthy soil contain air, are filled with water.

Savanna: The term applied to a type of tropical climate, with a wet and dry season, in which the most common form of vegetation is the tall tropical grass 'savanna'.

Scale: The name given to many kinds of small, usually dry leaves or bracts.

Scion: A bud or branch cut for grafting or planting.

Scion bank: A place where scion materials are stored for a certain period before they are used for grafting.

Scouring or pearing (Rice): Process of bran removal locally known as polishing.

Seasonal consumptive use: The total amount of water consumed in evaporation and respiration by a crop during the entire growing season expressed in depth of volume of water per hectare.

Seasonal drought: A kind of drought which occurs in during distinct annual period of dry weather.

Secondary dormancy: Dormancy that develops within the moist seed after it is removed from plant and is subjected to adverse environmental conditions.

Secondary fertiliser plants: Calcium, magnesium and sulphur, which are next in important to major nutrient for optimum crop growth.

Secondary growth: Growth, which results from the activity of vascular, or cork cambium in the formation of secondary tissues and mainly resulting in an increase in diameter or stem girth of a plant/tree.

Secondary plant nutrient: The secondary plant nutrient are calcium, magnesium and sulphur and are called secondary because these are not applied as straight commercial fertilisers.

Secondary productivity: The productivity of pasture/grassland measured in terms of animal grains/products such as milk, meat or wool instead of herbage yield is called as 'Secondary Productivity'. This is an indirect measurement of pasture productivity.

Secondary tillage: The tillage operations following primary tillage to create a good seed bed for proper seeding/planting.

Sectoral chimera: A chimera where genetically different tissues situated side by side in the same plant part.

Sedge: Plant belonging to the family cyperaceae, a large family of monocotyledonous plants distinguished chiefly by having achenes solid stems and three ranked stem leaves.

Seduction: In bacteria; the incorporation of bacterial genes into a sex factor (*e.g., F*) and their subsequent transfer to other bacterial cells through conjugation.

Seed: The mature ovule consisting of kernel and its integuments.

Seed garden: A plantation usually of a superior race raised or treated for the production of seed.

Seed orchard: Orchards that are established specially for seed production.

Seed rate: Seed rate is the quantity of seeds, measured either in volume or by weight, to be sown over an area. It usually expressed in kg per hectare. Optimum seed rates are recommended by State Agricultural Universities and Research Stations.

Seed selection: Using seeds of a good quality is one for the most important requirement for higher yield. Traditionally, female members of a farm collect the preferred types of seeds from the spikes before the bulk harvest, and also preserve them separately. With the advent of high yielding or hybrid seed, this practice is gradually declining. The seed must be procured from approved seed producing agencies like the Seed Development Corporation or the National Seed Corporation. The chosen seed should have high viability or a good germination percentage. It should not be infested with any pests and disease.

Seedbed: Soil prepared by natural or artificial means to promote the placement and germination and seed and growth of seedlings.

Seedling: A young tree generally less than 1-inch diameter; young trees that are still in the nurseries before planting.

Segregation: Separation of the two alleles of a gene during gamete formation and their passage into different gametes; usually due to the separation of homologous chromosomes during A I; each gamete contains only a single copy of only one allele; and leads to the typical 3 : 1 or 1 : 2 : 1 ratios in the F_2 generations of monohybrid crosses.

Segregation: The process of separation of the paternal and maternal chromosomes at meiosis.

Segregation distorter: A gene or some other genetic element which alters the ratio of segregating alleles recovered in the gametes; in *Drosophila*, the *SD*+ cells are eliminated during gametogenesis.

Selection: The process of allowing only a certain group of individuals from a population to become the parents of next generation.

Selective herbicide: It is that which is more toxic to one plant, some may be killed and others may be affected only slightly or not at all.

Selectivity: Property of differential tolerance; essential for the proper functioning of herbicides that kill some plant species when applied to a mixed population, but without serious damage to other species.

Self-compatibility: A condition in which the male gametes are capable of fertilizing the female gametes of the same flower.

Self-fertilization: The union between male gametes of an individual with the female gametes produced by the same individual; usually in plants, *self-pollination*.

Self–incompatibility: A condition in wl ich the male gametes are incapable of fertilizing the female gametes of the same flower owing to genetically controlled physiological hindrances.

Self-pollination: The transfer of pollen from a flower to the stigma of the same flower or another flower of the same plant.

Self-sterile: Inability of a cultivar to produce fruit with viable seeds following self-pollination because of non-functional male or female gametes.

Self-unfruitful: Inability of a cultivar to produce commercial fruit crop following self-pollination and parthenocarpy.

Semi-arid: A climate with average annual rainfall of about 200-900 mm with high variability of rainfall.

Semi-arid zones: Zones delineated by Thornthwite's moisture indicate varying from –20 to 40. These indices are defined by 100s – 60 D/n; where, s = means annual water surplus; D = means annual water deficiency; n = means annual water need.

Seminars: Seminar is one of the group discussion methods where a different aspect of a particular subject is discussed is a logical sequence under the guidance of experts and experienced professionals.

Senescence: The last stage of development when anabolic biochemical processes give way to catabolic processes leading to death of the tissue.

Sequence cropping: In this system, harvesting of the first crop is followed by the planting of the second crop, after fresh land preparation. Sequence cropping is adopted when water resources are adequately available till the maturing of the crop that is grown after the first crop. Sequence crops that are ideal after rice are mustard, pulses, maize, tomato, groundnut and vegetables.

Sequencing of contents: The order in which contents of a training programme are presented to the participants.

Sequential cropping: Growing two or more crops after each other in the same field per year. The succeeding crop is planted after the preceding crop has been harvested.

Sericulture: The technique of silk worm rearing for silk production. An agro-based cottage industry. It combines the agricultural base of mulberry cultivation and silkworm rearing with the labour intensive industrial superstructure involved in silk reeling, twisting and processing. One hectare under mulberry can employ 12 persons directly or indirectly right from cultivation upto silk weaving. A farmer with 0.4 ha of land can earn Rs. 15000 a year from sericulture with proper land management. It, thus, is an important alternate land use for marginal rainfed farmers. Silkworm seed can be procured from the Government. The Central Silk Board, State Sericulture Departments or private agencies in the State, provide subsidies for mulberry cultivation and for rearing silkworms. Mulberry is easy to grow because it can be grown in all agro-climatic zones and all types of soils with good drainage. Even slightly acidic or alkali soils can be treated with gypsum or lime and cultivated. It only needs some protective irrigation. For best results mulberry should be grown in rows across the slope.

Sessile: Not stalked.

Sex chromosome: A chromosome which differs in number or kind between the male and female individuals of the same species, and is involved in sex determination. (*Syn.*, allsome).

Sex factor: *See, F Factor.*

Sex mosaic: *See, Gynandromorph.*

Sex reversal: A change in the characteristics of the individual from male to female and *vice versa.*

Sex-influenced trait: Alleles of the gene governing such a trait show opposite dominance relationship in the males and females of the species.

Sex-limited character: A character which is expressed in one of the two sexes only.

Sex-linkage: The association between a character and the sex during inheritance; the concerned gene is located in the chromosome; in human beings several genetic diseases are sex-linked.

Sexual reproduction: The production of new individuals through the fusion of male and female gametes.

Sharing: The process in microlab in which the participants are encouraged to share an aspect of his or her life with others.

Shelter belt: It is a belt of trees and shrubs intended to protect field crop from wind and to conserve soil moisture. A windbreak is also a barrier for protection from winds but is most commonly associated with mechanical or vegetative barriers for buildings, gardens or opens. A good shelterbelt reduces wind velocity on the windward side from 5 to 10 times the height of the trees and on the leeward side from 10-30 times its height. Trees about 50 feet in height would protect an area approximately 400 m in width. The shelterbelt should consist of several rows of trees and trees of varying heights. A high row of shrubs on the windward side can provide a dense barrier to the wind. This arrangement ensures that the foliage of the shelterbelt lifts the wind well above ground level and gradually drops it downward on the other side of the shelterbelt. To have protection from the wind at all seasons both evergreen and deciduous trees are included in the shelterbelt.

Shifting cultivation: A form of agriculture in which soil fertility is maintained by rotating fields rather than crops. A piece of land is cropped until the soil shows sign of exhausting or is overrun by weeds, when the land is left to regenerate naturally while cultivation in done elsewhere. Now sites are usually cleared by firing (slash and burn).

Shoot: A branch, stem or twig.

Short wave radiation: Radiation with wavelengths less than 4 microns.

Short-day plants: The plants that normally develop flower with photoperiod (day light period) of 8-10 hours, *i.e.*, less than the critical day length.

Shot berry: Underdeveloped seedless berry of grape that fails to enlarge, caused due to boron deficiency or poor pollination.

Shoulder bund: It is a bund which runs along the entire outer edge of the terrace. It stops water from running off the side of a bench and thus prevents the terrace from collapsing. It is usually 30 cm high, with a bottom width of 75 cm and a side slope of the ratio 1:1. Shoulder bunds are also found along the outer margins of waterway.

Shower: In weather reports, solid or liquid precipation from a convective cloud is designated a shower and is distinguished in such reports from the precipitation intermittent of continuous, from layer clouds. Showers are often characterized by short duration and rapid fluctuations of intensity. Hail invariably implies a shower, while drizzle very seldom does.

Sib-mating: Mating between individuals which are progeny of the same two parents. (*Syn.*, brother-sister mating).

Silage: Animal feed resulting from the storage and fermentation of green or wet crop under anaerobic conditions.

Silage crop: Those crops, which are preserved in a succulent condition by partial fermentation in a tight recepticle.

Sile: A semi-airtight structure designed for use in the reduction and storage of silage.

Silvihorticulture: Care and cultivation of small plots (gardens) devoted primarily to vegetables and fruits, including woody spices.

Silvipastoralism: Agroforestry system in which livestock are fed the forage produced by trees and shrubs, and graze shorter plants such as grasses and herbs.

Silvipasture: This is a land management system in which trees and grasses are grown on the same piece of land. Forage, either shrubby or herbaceous is grown as the under storey crop with a mixture of top feed species of forest trees. Silvipasture has been found to be a dependable proposition for the utilisation of degraded lands in erratic rainfall areas because trees can tolerate extreme soil and climatic conditions while grasses and legumes provide good land cover. The system is most preferable where there is fodder shortage. **Advantages of Silvipasture:** Suitable grass-legume and tree combination *stabilizes* soil fertility and soil structure. Provides nutritious fodder for livestock without decreasing growth and yields of trees. In monocrop plantations there is no income till it comes to production. In Silvipasture system, pasture component can be utilized right from first years of its establishment. Grass cover keeps off weeds. Trees and grasses do not compete for soil moisture and nutrients. Root zone of grasses is limited to about 20 cm of topsoil; trees draw nutrients from deeper zones. With better moisture regimes in the system, the foilage will be available for a longer period in a season. Top field from fodder trees can be conveniently used during lean month, increasing carrying capacity of land.

Simulation methods: Simulation method consists of simulating real life experiences.

Single cross: A cross between two inbred lines, clones, cultivars etc.

Sink: The site in a plant with a high rate of metabolic activity where food resources are used.

Skill-gap analysis: The skill-gap analysis is a process of determining the training needs of individual employees in relation to the components of tasks identified.

Slope percentage: It is the difference in elevation between the highest point and the lowest point. It is also refered to as the dip to the land. The slope is expressed as fall per 100 metres or feet.

Sloughing of citrus: A disorder of pink and red grapefruit characterised by progressive browning and disintegration of rind which remains soft and moist and then falls away or slough off leaving the flesh exposed.

Slow release fertilizer: Fertilizer whose nutrients are present as a chemical compound or in a physical state that their availability to plants is spread over a period of time.

Smokeless *chullah*: The smokeless *chullah* (stove) is an improvement over the traditional *chullah*, both in terms of thermal efficiency (upto 20-25 per cent as against the 5-10 per cent efficiency of traditional *chullah*) and overall wood consumption (if properly used, they can save upto 1000 kg per year). There are several other advantages of this system such as reduced cooking time and release of smoke through a *chimney* (this is less harmful to health, and eliminates the need for fanning or blowing). A small-scale unit for manufacturing smokeless *chullahs* can be set up in a watershed as it requires only mud and skilled or semi-skilled labour. This provides employment in pottery and ceramics. The department of Non-conventional Energy Sources has implemented a programme of smokeless improved *chullahs* at grassroots level through state government agencies and other bodies such as Housing Corporations, Khadi and Village Industries Commission, National Dairy Development Board and Voluntary Organisations.

Smothering crops: Those crops which are through their thick canopy exclude light and prevent all top growth of weeds.

Smudging: Building up of slow fires emitting smoke in the orchard which induce flowering in some fruit trees as smoke contains acetylene.

Sod: Top few centimetres of soil premeated by and help together with grassroots of grass-legume roots.

Sod culture: An orchard management practice on slopy land in which vegetative cover is established by growing vines, legumes or grasses to prevent serious soil erosion.

Sodium adsorption ratio (SAR): A ratio for soil extracts and irrigation waters used to express the relative activity of sodium ions in exchange reactions with soil.

Soft scald of apple: A low temperature disorder where symptoms appear as smooth, brown, irregular shaped well defined areas of skin irrespective of the skin colour but usually not at the calyx end.

Soft water: Water which does not contain those minerals that prevent free lathering when soap is added.

Soft wood grafting: In this method stock are raised *in situ* for one year or more and grafting of scion by wedge grafting method on terminal new growth of the stock.

Soil: Soil is the unconsolidated mineral material on the immediate surface of the earth. It is a dynamic natural body composed of mineral and organic materials and living forms in which plants grow. It is formed by the integrated effect of climate and living matter acting upon parent material over time. It takes around 400 years to form a 2-3 cm layer of top soil. Every year in India, we lose 6 billion tonnes of rich top soil through erosion with the result that almost 50 per cent of the land is in some state of degradation.

Soil amendment: Any substance that is added to the soil for the purpose of improving its physical or chemical character, enhancing soil productivity or promoting the growth of crops but excluding commercial fertilisers and organic manure.

Soil borne diseases: These diseases usually affect seeds and seedlings. Disinfecting and sterilizing soil helps to destroy the pathogens. Treatment includes broadcasting chemicals into soil either in furrows or at random. Fumigation and drenching of soil are other methods.

Soil colloids: The colloidal fraction of the soil primarily made up of inorganic material (clay) with varying amounts of organic colloids (human).

Soil conservation: A combination of all management and land use methods, which safeguard the soil against depletion or deterioration by natural or by man-induced factors.

Soil consistency: It is the capacity of the soil to resist manifestation of the physical forces of cohesion and adhesion, acting within the soil at various moisture contents.

Soil cultivation: Tillage operations performed to create soil conditions conducive to improve aeration, infiltration, compaction and moisture conservation as well as to control weeds.

Soil cutting: Soil separation from the soil mass by slicing action.

Soil failure: The alternation or destruction of a soil structure caused by mechanical forces, such as shearn compression and impact.

Soil incorporation: Mechanical mixing of herbicides in the soil.

Soil injection: Placement of herbicide beneath the soil, with minimum of stirring or mixing of the soil. Injection is done by an injection blade, knife or line.

Soil moisture: The moisture content of soil is generally as the percentage ratio of the mass of water to that of dry soil, but may be expressed also in terms of inches of water at given depth of soil.

Soil moisture deficit: The amount of rainfall or irrigation required to restore soil to its field capacity.

Soil moisture regimes: The soil depth upto which moisture is held after irrigation or rainfall should coincide with the depth of the root zone to provide adequate moisture for better crop growth. This determines the irrigation interval or the frequency at which the field should be irrigated.

Soil moisture stress: It is the sum of soil moisture tension and the osmotic pressure of the soil solution.

Soil moisture tension: It is the force with which the water held by soil against gravity. It is generally expressed in mm of mercury.

Soil persistence: Refers to the length of time that an herbicide remains effective in the soil and exhibits some degree of phytotoxicity to some plant species.

Soil *p*H: Soil acidity or alkalinity.

Soil productivity: It is the present capacity of a soil to produce crop yield under a defined set of management practices. It is measured in terms of the yield in relation to the inputs of production factors.

Soil profile: An examination of a vertical cross section of soil from the surface through all the sections reveals the presence of more or less distinct horizontal layers. Such a section is called the profile and individual layers are called horizons. Each layer is distinguished from the other by its morphological characteristics like colour, texture, structure etc. Generally the profile consists of horizons A, B, and C.

Soil salinity: The amount of soluble salts in a soil, expressed in terms of percentage, parts per million (ppm), or other convenient ratio.

Soil science: That science dealing with solid as a natural resources on the surface of the earth including soil formation classification and mapping, and the physical chemical and these properties in relation to their management for crop production.

Soil separates: Individual soil particles after standard dispersion treatment recognised by the International Society of Soil Science are (*a*) coarse sand –2.0 to 0.2 mm, (*b*) fine sand –0.2 to 0.02 mm, (*c*) silt –0.02 to 0.002 mm, and (*d*) clays < 0.002 mm diameter.

Soil solarization: The process of heating the surface soil by raising the soil temperatures.

Soil Sterilant: Herbicide which renders the soil incapable of supporting growth of all plants; sterilization may be temporary (a few months) or relatively permanent (years).

Soil structure: The combination or arrangement of primary soil particles into secondary particles, or groups of aggregates.

Soil texture: It is an expression of the distribution of the various particles size present in soil is described as a course; medium or fine textured depending on the predominant particle size.

Soil-moisture characteristic curve moisture retention (curve): A graph showing the relationship between the amount of water remaining in the soil at equilibrium as a function of matric suction.

Solar cookers: Suitable for baking, cooking and boiling. Process takes less than 150 minutes in even cold weather. Can also be used as a hot case as it retains heat for five hours after cooking.

Solar farm: A suggested power-utility based in a desert and conferring a considerable area, where large amounts of electrical energy would be generated from solar energy by capturing it by photovoltaic devices.

Solar heating system: Heats water at low cost and can be used easily in homes and industries.

Solar photovoltaic panels: Can provide power to rural areas and no fuel needed for generation of power.

Solar pumps: Used to pump irrigation water and can save electricity. Are effective for drawing water upto 7-8 mts high.

Solar stills: This is used to convert saltish or brackish water to potable water. The technique used is very simple and thus it can be manufactured with locally available material.

Sole cropping: Growing one crop variety in a pure stand.

Sole crops: A crop grown in pure stand at optimum population (spacing).

Solubility: A measure of the amount of substance that will dessolve in a given amount of another substance.

Soluble sodium percentage (SSP): The proportion of sodium ion in solution in relation to the total cation concentration.

Solution concentrate: A liquid formulation, also known as soluble concentrate, that consists of a herbicide dissolved in a solvent system, which forms a solution when added to water or a carrier.

Somatic cell: A cell making up the body of an organism, excluding the germ cell; such cells have the full somatic complement of the species.

Somatic chromosome complement: The number and the type of chromosomes normally found in the somatic cells, particularly in the meristematic cells, of a species, it is twice the gametic complement (n) and is denoted by $2n$.

Somatic chromosome number: The chromosome number normally found in the somatic cells, more specifically in the meristematic cells, of a species; represented by $2n$.

Somatic *hybridization*: Fusion of protoplasts of the same or different genotypes which combine two nuclei and two cytoplasm leading to the development, of somatic or parasexual hybrid.

Somatic mutation: A mutation in somatic cell, usually in meristematic cell. (*Syn.*, bud mutation).

Somatoplastic: The collapse of a fertilized ovule during the early development stages due to disturbed embryo-endosperm ratios.

Sorosis: A multiple fruit developing from a spoke or spadix where the flower fuse together by their succulent sepals and at the same time the axis bearing them grows and becomes fleshy or woody. And as a result the whole inflorescence forms a compact mass.

Sorption: It refers to surface-induced removal of a chemical substance from solution by adsorption, absorption or precipitation.

Sowing: The process of spacing seeds in seed bed.

Spacing: The distance between crop rows (inter-row spacing) and between plants within the row (intra-row spacing).

Spade leaf: The first true leaf of a seeding clover or Lucerne. The leaf is simple in contrast to the subsequent trifoliate leaves.

Spadix: A spike with a fleshy axis which is enclosed by one or more large, often brightly colored bracts, called spathes.

Spatial arrangement: It is defined as the pattern of distribution of plants over the ground, which determined the shape of the area available to the individual plant.

Spawn: The production of eggs by fish, frogs, shellfish and other animals growing or living in water.

Specialised cropping: It is a cropping plan in which a single crop contributes 50 per cent or more of the total production or monetary receipt (comparable equivalent) in one year.

Species: A group of interbreeding or potentially interbreeding individuals which is reproductively isolated from other such groups.

Species: The unit of classification into which genera are sub-divided.

Specific crop intensity index: Specific intensity index determines crop or group of crops compared to total time available to the farmer for crop production during the time period under study.

Specific gravity: Ratio of the weight of any volume of substance to the weight of the same volume of water.

Specific humidity: The ratio of mass of water vapour in a volume of moist air to the total mass of the volume of moist air.

Specific leaf area: The leaf area per unit leaf weight expressed as cm^2g^{-1} and is given by, Specific leaf area = Area of the leaf (cm^2)/Weight of the leaf (g).

Spectrophotometer: Photometer adopted to measure the spectral distribution of the luminous flux in a beam of light.

Sperm: A mature male gamete. (*Abbreviation* of Spermatozoon; *pl.,* Spermatozoa.)

Spermatogenesis: The process of production of mature male gametes or sperms from diploid sperm mother cells through meiosis; in case of animals.

Spike: The inflorescence in which the main axis is elongated and the lower flowers open earlier than the upper ones in raceme, but the flowers are sessile. Example is mulberry.

Spikelet: It is the unit of the panicle.

Spillman's equation: According to Spilman plant growth increase with increasing application of limiting element but increase in growth is not in direct proportion to the amount factor added. The increase in growth with each successive addition of the element in question is progressively smaller.

Spillway, chute: It is generally used to drop water at reaches much farther than a drop structure. It can be used to control gully drops upto 6 metres. It is also much cheaper to install when compared to a drop spillway or a drop inlet spillway. There is however the danger of it being undermined by rodents as they attack the gully banks. In locations with poorly drained soils, the foundation of a chute spillway may be threatened by seepage.

Spillway, drop: This is an overfall structure where water drops over a vertical wall into an apron. It is constructed across the channel and the runoff water, flows through a weir across the apron. The energy of the water thus gets dissipated. Drop spillways are generally adopted for stabilising drops upto 3 metres.

Spillway, drop inlet: This is an overall structure where water drops through a vertical riser into a discharge conduit. It is used as an outlet for farm ponds or reservoirs and is constructed along the embankments. In gully control works, a drop spillway is used in combination with earthen bunds when there is a fall of more than 3 metres into the gully bed.

Spillway, emergency: This is provided in farm ponds and reservoirs to safely handle the peak flows. It is usually located above the mechanical spillways.

Spillways: Spillways are channels for conveying excess water from reservoirs. They can be open or closed. They contain manually or automatically operated gates or regulate the flow of excess water. The spillways include chute spillways, drop spillways, drop inlet spillways and emergency spillways.

Spongy tissue: A serious disorder of Alphonsa mango, in which fruit develops bad odour with sponge like mesocarpic tissue having air pocket due to inactivation of hydrolytic enzyme by connective heat.

Spontaneous mutation: A mutation that arises naturally without the application of any mutagen (by man); the cause of such mutations is not definitely known.

Sporogenesis: The production of haploid micro- and megaspores from the diploid microspore and megaspore mother cells, respectively, through meiosis in plants.

Sporophytic incompatibility: In this type of incompatibility, reaction is controlled by the genotype of the pollen producing plant, there are smaller number of alleles involved.

Spot treatment: It is the term used to the application of weed killers to individual weeds or of a restricted area to control an infestation of a weed species requiring special treatment.

Spray drift: The movement of airborne spray particles from the spray nozzle beyond the intended contact area.

Sprays and dusts: A weed killer as spray which is in the form of a liquid, solution or a suspension capable of flowing in fine drops and as dust when it is used in the form of a finally divided powder.

Spreader sticker: A surfactant closely related to wetting agents which facilitates spreading and increases sticking of a herbicide on vegetation, *e.g.*, tenax.

Spreading agent: A substance used to improve wetting, speading or possibly the adhesive properties of an herbicide spray solution.

Spur: Short, woody stem which is the principle fruiting area of most of the fruit.

Squall: A strong wind that rises suddenly, generally lasts for some minutes, and dies comparatively suddenly away. It is distinguished from a gust by its longer duration.

Stability: Collective aspect of system characteristics that minimise the negative effects of abrupt, unexpected change on the farm system.

Stalon: Common in grasses, this slender branch or shoot is capable of forming both buds and roots and is, thus, used in plant propagation.

Stamen: The male reproductive parts of a flower.

Staminal carpellody: The transformation of stamens into fleshy carpel like structure during the early development of a flower, e.g. papaya.

Staminate flower: A flower bearing stamens only.

Staminode: A structure appearing in place of a stamen but bearing no pollen.

Standard deviation: The square root of variance; a measure of the amount of variation present in a population.

Standard error: A measure of variability in a population of sample means derived from a single population; an indicator of the correspondence between sample estimates and population parameters; estimated from a sample as $\sqrt{\text{variance}/n}$.

Standard meterological week: All the rainfall data is represented in a week commencing from the first of June and recorded thoughout the fifty-two weeks of the year. The precipitation received during the week is taken into consideration to record along with temperature so as to find out the effect on yield and behaviour of crops. For the purposes of rainfall, the first week in January is denoted as the first standard meteorological week and the last week in December is denoted as the 52nd standard meterological week.

Standard rain gauges: collection of rainfall through an opening of regular dimensions and through this work out the amount of rainfall.

Standards: This refers to the acceptable level of performance in terms of quality and quantity.

Starvation tests: In biological control of weeds with the help of insects, careful feeding tests (also called starvation tests) are carried on before the insect is released in new locality to prove the limited feeding habits.

Statistic: An estimate of a population parameter, *e.g.*, population mean, obtained from a sample drawn from the concerned population.

Stem: The main axis of a plant, usually above ground and supporting leaves, flowers and fruits.

Stem foliage application: An application of an herbicide to both stems and foliage of a plant.

Sterile: Incapable of sexual reproduction.

Sterility: Inability to produce progeny usually due to non-functional gametes; male and female sterility.

Stigma: The terminal part of the pistil that receives the pollen.

Stolon: The above-ground runner or stem that develops roots and shoots at the tip or nodes (*e.g.*, *Cynodom dactylon,* bermuda grass).

Stone fruits: Fruits; also known as "drupes" with one or more seeds.

Stooling: The act of cutting off a stump or root of a plant to allow young shoots to grow.

Storage rain gauges: They are used in remote and inaccessible areas where it is not practicable to take observations everyday or every week. These gauges have larger capacities and observations may be taken after a month, season or year.

Straight fertiliser: Qualification generally given to a nitrogen, phosphorus or potassium fertiliser having a declarable content of one primary nutrient only.

Strain: An experimental line.

Stratification: A method of breaking seed dormancy in which the imbibed seed are subjected to a period of chilling to after ripen the embryo.

Stratus: One of the cloud genera, generally gray cloud layer with a fairly uniform base, which may give drizzle ice prisms or snow grains. When the sun is visible through the cloud, its outline is clearly discernible. Stratus does not produce halo phenomena except, possibly, at a very low temperature.

Straw or stubble mulches: The dried remnants of crop plants, from which the seed has been removed, also called stover. Includes mulches of straw, stubble and other crop residues. When left on the soil surface, they act as a barrier to the absorption of solar energy by the soil and reduce the velocity of air movement above the soil. They thus help reduce evaporation loss from the soil surface. Plastic mulches also serve the same purpose. These mulches provide an added advantage of increasing the soil temperature, hastening germination of new crops like wheat and barley. They are, however, very expensive.

Streamline: A curve which is parallel to the instantaneous direction of the wind vector at all points along it. Isobars are streamlines only in strict geostrophic flow.

Stress: A potentially injurious force or pressure acting on the plant which may lead injury or death.

Strip cropping: Growing soil conserving and soil depleting crops in alternate strips running perpendicular to the slope strips wide enough to permit independent cultivation but narrow enough for the crop to interact agronomically.

Strip plot design: Two different sets of treatments can be tried in large plot with one set of plots superimposed over the other set at right angles. A block may be divided into strip in one direction to be allotted to one set of treatments and into another set of strips to be allotted to the second set of treatments. This design is used when two sets of treatments need larger plots.

Structural gene: A gene which codes for (through mRNA) the amino acid sequence of a polypeptide.

Structured experience: A structured experience is a group learning design which is based on the five stages of experiential learning cycle which include experiencing, publishing, processing, generalizing and applying.

Stubble: The basal portion of the stems of plant left standing after cutting or harvesting the crop.

Stubble mulching: Stirring the soil with implements that leaves considerable part of the vegetative material or crop residues or vegetable litter on the surface as a protection against erosion and for conserving moisture by favouring infiltration and reducing evaporation.

Style: The more or less elongated part of the pistil between the ovary and the stigma.

Sub-humid: In the tropics, a climate with average annual rainfall of roughly 900-1500 mm or region where the moisture contents is below that under humid condition but normally still adequate for the production of many crops without irrigation.

Sublethal gene: A gene which causes the death of most, but not all, of the individuals that carry it in the appropriate genotype.

Subsistance agriculture: Farming system in which a large part of final yield is consumed by the producer. Most subsistence system involve production of some crops or animals for sale, but the ratio of subsistence to cash production may vary greatly from year to year.

Subsistance farmings: It is farming enterprise, which provides food and commodities just sufficient for the farming family, and there is no surplus to sell.

Substitution cropping: It is substitution of an existing inefficient crop with an identified efficient crop in a zone or region.

Subvital: A gene which kills some of the individual which carry it in the appropriate genotype.

Succession: An orderly process of change in a community (of plants, animals, soil microbes, etc.) that results from modification from the environment by organisms and culminates in a system attaining a steady state, or climax.

Suckers: A shoot growing from an underground stem or root.

Sulphur coated urea (SCU): It has been developed in recent years as a slow release nitrogen fertiliser and is basically urea with a coating of element sulphur, including a binding agent, a sealant and a microbicide. The N content of SCU ranges from 10-37 per cent depending on the thickness of the sulphur coating.

Sunscald of apple: Bleached or brown areas on the side of the fruit exposed to the sun, sometimes accompanied by water soaked sub-epidermal tissue which often becoming darker after removal from storage, the cortex tissue under lesion is likely to be diffusely brown and bitter in taste.

Sunscald of pineapple: A disorder which occurs due to exposure of fruit direct sunshine when the peduncle bearing the fruit fall on one side where cells under the skin of exposed surface get damaged.

Sunshine hours: The number of hours of bright sun shine per day, also sometimes defined as the duration of traces of burns made on a chart by Campbell-Stokes recorder.

Sunshine recorder: An instrument for recording the duration of bright sunshine. Such instrument depends either on the heating action of the sun or on the chemical action produced by the sun rays.

Superfemale: *See, Metafemale.*

Supermale: A weak and almost completely sterile male individual exhibiting more intense secondary sex character than normal males; in *Drosophila*, due to an excess of autosomal sets in relation to the number of *X* chromosomes as compared to that in the normal males; ($X/A = 0.5$ = male; $X/A < 0.5$ = supermale).

Suppressor mutation: A mutation which suppresses the phenotypic expression of another mutation; usually, in another gene, or in the same gene at a different site.

Surface tension: A physical property of liquids, due to molecular forces, that causes them to form frops, rather than spread as a film.

Surfactant: A surfactant is a material that exhibits activity at surface or interfaces, *e.g.*, detergents, wetting agents, spreaders and sticking agents.

Suspension: Liquid or gas in which very fine solid particles are dispersed but not dissolved.

Sustainable agriculture: Management of resources for agriculture to satisfy changing human needs, while maintaining or enhancing the quality of the environment and conserving the natural resources.

Sword sucker: The sucker of banana which have well developed base with narrow sword shaped leaf blades at the early stages.

Syconus: A fruit type which is developed from a hollow, pear shaped, fleshy receptacle enclosing a number of minute, male and female flowers and the fleshy receptacle grows to form the fruit.

Symbiont: An organism living in a close association with another organism in such a way that both of them benefit from the association.

Symbiosis: The relationship of two or more different organisms in a close association that is beneficial to each organism.

Symplast: Functionally integrated unit consisting of all living cells of a multi-cellular plant; all living cells connected by plasmodesmata; including the phloem.

Sympodial: A composite axis produced and increased in length by a successive, development of lateral buds just behind the apex.

Symposium: Symposium is a gathering of experts who present their findings or ideas on a specific sub-topic related to a major subject.

Synapsis: During prophase I of meiosis; close, longitudinal, point-to-point precise association between homologous chromosomes. (*Syn.,* Chromosome pairing).

Synaptenemal Complex: A proteinaceous structure about 1000 A wide formed between homologous chromosomes during synapsis; considered to facilitate recombination or crossing over.

Syncarpous: A condition in which the carpel are fused.

Syndicate: A syndicate is a group of people (8-10 in number) assigned with a task of investigating a particular problem or topic and to find out the solution.

Syndrome: A group of symptoms which occur together and characterise a particular disease.

Synergism: Cooperative action of different chemicals so that the total effect is greater than the sum of the different effects.

Synergy: The action of two or more substances, organs or organisms to achieve an effect of which is individually capable.

Synoptic chart: A chart or map on which is represented the distribution of selected meteorological elements over a large area at a specified instant of time. 'Surface synoptic chart'is an alternative for weather map.

Synoptic meteorology: That branch of meteorology which is concerned with a description of current weather as represented on geographical charts and applied especially to the prediction of its future development.

Synoptic station: A station at which meteorological observations are made at the 'major synoptic hours' 00. 06. 12 GMT. The observed elements being plotted symbolically on the weather map.

Synteny: Two or more genes being located in the same chromosome without any reference to their linkage relationship. (*Adj.*, Syntenic).

Synthetic: A cultivars produced by intercrossing number of selected genotypes and maintained by open-pollination.

System approach: An approach for studying a system as an enmity made up all of its components and their relationships, together with relationships between the system and its environment.

Systematic herbicide: A herbicide which is translocated within the plants.

T

Tap root: A root originating from the radical in dicot plant having a prominent central portion growing vertically downward and giving off small lateral roots in succession.

Tapping: The process of harvesting latex from trunk of rubber plant, fruit of papaya, etc.

Task analysis: Task analysis is a process, by which one can know the different elements or sub-tasks involved in a task and also to find out the specific elements of the task, which are critical for its performance.

Task identification: Task identification refers to dividing a major work event into its component parts, which allows a trainer to understand what an individual does in the course of his work.

Tassel (maize): The flower cluster at the tip of corn plant comprised of pollen-hearing flowers, *i.e.*, the staminate inflorescence of maize.

Teaching: Teaching refers to the process of imparting knowledge in typical classroom situations where the teacher controls the process and contents of learning.

Technology: The combination of knowledge, inputs and management practices, which are developed together with productive resources to produce a desired output.

Telocentric: A chromosome having centromere at one end (in the place of telomere).

Telophase: The last stage of cell division following anaphase preceding interphase; the chromosomes gather at the opposite poles and being to uncoil to finally form loosely packed masses of chromatin fibres.

Teminalization: More commonly, *chiasma terminalization*; the movement of visible chaismata toward the ends of chromosome arms due to the movement away from each other of the homologous chromosomes in each bivalent; the movement is more pronounced at the centromeres; terminalization begins in the diplotene of prophase I of meiosis and ends at the onset of diakinesis.

Temperate phage: A phage which may integrate into the chromosome of its host and may remain there as a prophase (*lysogeny*), or may dissociate from the chromosome of its host and lyse it like a virulent phage.

Temperate zone: Either of two latitudinal zones on the earth's surface lies between 23°27″ and 66°27″ north and south.

Temperature inversion: An increase of air temperature with height so that warmer air overlies colder air.

Template: A mould; a substance used to create identical copies as specified by itself; DNA molecules serve as templates during replication and transcription.

Tendril: A slender, simple or branched organ arising from leaf or stem or part of them which can twine round a support.

Tensiometer: A device for measuring the negative pressure (or tension) of water in soil *in situ* consisting a porous permeable ceramic cup connected through a tube to a manometer or vacuum gauge. It is used in scheduling of irrigation.

Tenure: The right of property, granted by custom and/or law, which may include land, trees, and other plants, animals and water.

Terminal evaluation: Terminal evaluation is done at the end of the programme to find out to what extent the objectives and desired benefits of the programme have been achieved.

Terrace: An embankment or ridge of earth constructed across a slope to control runoff and minimise soil erosion. Bench terraces and ridge terraces are the two general types.

Terrestrial radiation: The long wave radiation emitted by the earth and atmosphere in the approximate temperature range 200-300K. The radiation is confined within the wavelengths of about 3 and 100 mm and has maximum intensity at about 10 mm.

Terrestrial weeds: Weeds, which normally start and complete their life cycle on land.

Test cross: A cross of the F_1 with the recessive parent.

Test plot: An area (usually about 3 ha) of a field where farmers agree to test the new agricultural technologies.

Testa: The outer covering of a seed.

Testcross: The cross of an F_1 hybrid with an individual or strain having the recessive phenotype for the concerned trait.

Tetrad: The group of four daughter cells produced by meiotic division of a single cell; also, the four-standard structure (*i.e.*, bivalent) produced due to the synapsis of two homologous chromosomes, each having two chromatids.

Tetraploid: A cell or an organism having four genomes; the genomes may be copies of a single genome (autotetraploid) or of different genomes (*allotetraploid*).

Tetrasomic: A cell or an organism having one chromosome pair in addition to the normal somatic complement of the species; one chromosome of the genome has four copies or homologous.

Thermal capacity: The thermal capacity, also called heat capacity, of a body is the product of its mass and its specific heat.

Thermo isopleths: Lines of equal soil temperatures.

Thermograph: Apparatus used to give a continuous record as a graphical chart of temperature with time.

Thermo-hydrograph: Instrument resulting from the combination of hydrograph and thermograph. It furnishes on the same chart simultaneously continuous recording of atmospheric temperature and humidity.

Thermoperiodicity: Effect of temperature difference between light and dark periods upon plant.

Thickness: The geopotential height difference at a given place between specified pressure levels. Thickness values relating to select standard pressure levels are obtained from radiosonde observations and are plotted on geographical thickness charts. Contours are drawn at an appropriate thickness interval, joining places of equal thickness and are termed 'thickness lines'. Analysis of such charts is termed as 'thickness analysis' and has an important role in synoptic meteorology.

Thinning: Removing of extra plants from thickly populated crop stand with an idea of maintaining optimum plant population.

Three way cross: A cross between a single cross and an inbred (A × B) × C.

Threshold: The concentration, level or intensity below which a physical or chemical agent does not produce its effect.

Thunder storm: One or more sudden electrical discharges manifested by a flash of light (lightning) and a sharp or rumbling sound (thunder).

Tide ridging: Leaving or forming short ridges of earth at right angles to contour ridges so as to retain rainwater in the field.

Tillage: The operation of agriculture implements through the soil for the preparation of seed and root beds. It helps loosen soil, incorporate fertilizers and leads to fine lilth aggregates. Inadequate crop stand is a major reason for low crop yields in rainfed areas. This is mainly because of the restricted entry of water into the soil due to improper tillage practices. Tillage in rainfed areas agriculture therefore aims at—Preparing a suitable seed bed; Controlling weeds; Conserving moisture; Establishing uniform initial stand; Increasing water intake; Covering large areas during the restricted sowing period using lesser amount of energy; and maintaining the soil surface conditions. Land treatment works out to be cost-effective in the long run because proper tillage is a low cost investment. All one needs is a suitable implement and labour. There are different types of tillage practices for different soil types. They include deep tillage, off-season tillage and shallow tillage.

Tillage action: The action of a tillage tool in executing a specific form of soil manipulation such as soil cutting, pulverising or inversion.

Tillage, deep: This is used to break the hard pan that forms below the top soil.

Tillage, shallow: Tilling the soil upto a depth of 10 cm either with a peg tooth cultivator or a blade harrow is known as shallow tillage. It is normally conducted after the harvest of the crop.

Shallow tillage ensures moisture conservation through higher infiltration and results in low runoff.

Tiller: Grass stump arising from a lateral bud at the base of the plant is the process of tiller formation.

Tillering stage: The appearance the forest tiller from the axially bud in one of the lower lost nodes.

Tilth: The physical condition of the soil with respect to its fitness for the planting or growth of a crop.

Tip layering: A type of layering in which the growing tips of trailing type shoots are bent down and buried in the soil to the depth of 5-7.5cm to from roots.

Toddy: A sugar containing juice which is obtained by tapping the unopened spadix of coconut palm.

Tolerance: In the context of herbicide tolerance, it is defined as the natural or normal variability of response to herbicides that axists within a plant species and can easily and quickly evolve.

Top cross: A cross of an inbred and an open pollinated variety.

Top working: A grafting procedure employed primarily to change the inferior unwanted cultivar of the established plant either by grafting (top grafting) or by budding (top budding).

Topo climatology: Study of local climate as effected by slope and other surface characteristics.

Topography: The lay of the land indicating the relative positions of natural or man made features. The configuration includes altitude, slope and exposure. Information on topography is collected through land survey and is transcribed onto a topographical map which is a detailed and accurate physical description of a region. The configuration of surface including its relief and the position of its natural and man-made features.

Toposheet: It is a map of a particular area showing its topographical features. Usually contours are marked at 50 feet or 20 m vertical intervals. Toposheets reveal the position of land marks like forests, lakes, roads, temples, drainage courses, import land marks, churches, dykes and mountains. These can be obtained from the survey of India on requisition Training and Visit (T&V). A system of extension that focuses on the need for competent, well-informed village level extension workers who aid farmers in tackling their immediate problems by linking them with reasearch institutions. The methods uşed may vary from place to place to suit particular agricultural, social and administrative conditions. However, the essential features of continuous training and regular fixed visits by agricultural extension staff remain unchanged. The T&V system offers many advantages and if properly adopted can be successfully implemented under most conditions.

Topping (tobacco): The removal of the terminal bud with or without some of the small top leaves just before or after the emergence of a flower bud.

Total or Non-selective herbicide: Herbicides which will kill all or total vegetation present.

Totipotent cell: A cell capable of undergoing development like the zygote, that is, of developing into a normal embryo, under a suitable environment.

Trade winds: The trade winds (or tropical easterlies) are the winds which diverge from the subtropical high-pressure belts, centred at 300-4000 N and S, towards the equator, from north-east in the northern hemisphere and the south-east in the southern hemisphere.

Traditional agriculture: Farming system, which are based on indigenous knowledge and practices, and having evolved over many generations.

Training: Training is a process of acquisition of new skills, attitudes and knowledge in the context preparing for entry into a vocation or of improving ones productivity in an organization or enterprise. It refers to the practice of giving shape to the tree in the initial year of planting to bear the load of crop in subsequent year.

Training environment: The physical environment related to training such as size and shape of room, facilities related to electricity, size of the audience, lighting arrangement, and seating arrangements.

Training management: Training management is the process of planning, implementing, evaluation and follow-up of training programme so as to accomplish its objectives in an efficient way.

Training method: The term training method refers to a combination of instructional technologies, which are used in presenting the contents of the training to achieve the learning objective at the shortest time possible.

Training needs: Training need refers to the gap between 'what is' and 'what should be' in terms of incumbents 'knowledge, skills, attitudes and behaviour' for a particular situation at one point in time.

Trans-arrangement: *See, Repulsion Phase.*

Transcription: The production of an RNA molecule complementary to a DNA molecule which serves as the template for RNA polymerase, the enzyme that catalyzes transcription.

Transduction: Genetic recombination in bacteria following the transfer by a virus of a segment of the bacterial chromosome from one bacterial cell into another.

Transduction, abortive: The bacterial DNA transferred by a virus does not undergo recombination, fails to replicate and is gradually diluted and lost from the progeny cells.

Transduction, generalised: The phage may transfer any bacterial gene to the recipient cell; a phage particle contains a random segment of the bacterial chromosome; mediated by virulent phages.

Transduction, restricted: The phage transfer only those genes that are located near a specific site, *i.e.*, the site where a temperate phage integrates into the bacterial chromosome; mediated by temperate phages.

Transfection: The uptake of DNA by eukaryotic cells, followed by the incorporation of gene present in this DNA into the genome of eukaryotic cells.

Transfer RNA (tRNA): RNA molecules that transport specific amino acid to the ribosome, each tRNA species has a specific anticodon which base-pairs with the appropriate mRNA codon.

Transformation: In bacteria, genetic recombination following the uptake of a DNA molecule by a competent bacterial cell; in eukaryotes, the conversion of a cell growing *in vitro* to a state of uncontrolled growth similar to that of a tumour cell; in genetic engineering, introduction of a vector containing the desired gene into a host cell.

Transgenic: A term applied to organisms that have been altered by introducing DNA molecules into them.

Transgressive breeding: Isolation of recombinants possessing new characters or new intensity of traits surpassing both the parents in F2 onwards.

Transgressive segregation: The appearance of individual in the F_2 or a subsequent generation which exceed the parental limits with respect of one or more characters.

Transition: (*i*) The process of changing from one to another; refers here to the change of a farm system from either HEIA or LEIA to LEISA. (*ii*) The substitution, in a DNA or an RNA molecule, of purine by another purine or of a pyrimidine by another pyrimidine; leads to gene mutation.

Translation: Synthesis of a polypeptide in association with a ribosome and directed by an mRNA molecule with the help of tRNA and several highly specific enzymes.

Translocated graft incompatibility: Type of graft incompatibility in which the incompatibility condition is not overcome by the insertion of a mutually compatible inter-stock piece between the stock and scion.

Translocation: Integration of a segment of one chromosome into a non-homologous chromosome; this integration may be reciprocal, *i.e.*, the two non-homologous chromosomes involved in the translocation may exchange segments.

Transpirating: The process of planting seedlings in prepared seedbeds.

Transpiration: Transpiration is the process by which water vapour is released into the atmosphere through the stomal opening found on leaves and stem of plants.

Transpiration ratio: The quantity of water transpired to produce a unit amount or dry matter.

Transposon: A DNA segment which is capable of changing its location within a chromosome or between chromosomes.

Transversion: In DNA or RNA; substitution of a purine by a pyrimidine and *vice versa.*

Trap cropping: For the control of striga (agiya), which is a parasitic weed, a host crop of grass is planted which includes germination of weed seeds. The crop is then buried into the soil before the seed formation of weed. Such cereal or grass crops are known as "trap crops" which exert a cleanings effect on the land by inducing germination of number of Striga seeds. The practice is known as "trap cropping".

Tree cropping: It is based on the concept of using tree crops as the basis on an agricultural system. It entails replacement of grain crops with tree crops, consisting of an upper tree story of one or more types of forage trees, a ground level permanent pasture and one or more types of domestic animals that feed on the pasture or on tree products.

Tree patta scheme: This is a scheme under social forestry with two-fold objectives—to green areas; to provide employment and income to landless labourers. The government supplies planting material to the farmer and provides monetary incentive till the crop comes to maturity. Once the plants mature, the produce is shared between the government and the labourer. Trees in Silvipasture reduce fire in summer. *Trees Requirements*: Fast-growing with multipurpose uses, like fodder, fuel, timber etc. High palatability; Good coppicing ability; Ability to withstand browsing, lopping, intensive trampling; Resistant to drought and extreme temperatures. *Pasture requirements:* Ability to grow well even under shade; Tolerant to drought and extreme temperature; Easy propagation; High palatability, with good nutritive value; Ability to conserve soil water and nutrients; Ability to withstand over grazing and trampling.

Trellis: A frame of light strips of wood crossing one another with open spaces in between, especially one supporting growing vines.

Trench layering: Method of layering which consists of placing a stem in the shallow trench and cover with soil and after producing roots from the nodes, the new individual is removed from the mother plant.

Trench, contour: Contour trenches are dug along the contour and are used in areas with more than 600 mm annual rainfall. They help in conservation of scarce rainwater, encouraging it to infiltrate into the soil at all points. They also break the velocity of runoff, thus reducing erosion. Normally forest trees are planted on the berm with the distance between the trenches varying according to the tree canopy. If the canopy of the tree species extends upto 6 m a provision of 7 m space is made between two successive trenches.

Trench, graded: Graded trenches are used in areas with more than 600 mm annual rainfall. They are provided at a slight grade away from the contour. This ensures that excess water runs off safely and drains into a waterway. The primary purpose of graded trenches is therefore to drain excess water safely while conserving adequate amount as required.

Trench, staggered: The operation of agriculture implements through the soil for the preparation of seed beds. It helps loosen soil, incorporate fertilizers and properly deepens the sand bed. Inadequate crop stand is a major reason for low crop yield in rainfed areas. This is mainly because of the restricted entry of water into the soil due to improper tillage practices.

Trenching: Trenches are pits made across the slope to hold rainwater as it flows off. They are created along contours on land planned for forestry or horticulture and act as reservoirs recharging the ground water. There are two types of trenches.

Trickle irrigation: A method of applying water directly near the root zone of the plants through a number of flow-rate outlets generally placed at short intervals along small tubing. Sometimes it is referred as 'drip irrigation', followed for high value crops in arid regions.

Trihybrid: The F_1, F_2 etc. progeny from a cross involving two parents differing for three genes (or characters), or considered to be so for that study.

Triple cropping: Growing three crops a year in sequence.

Triploid: Having 3 x number of chromosomes.

Trisomic: A cell or an organism having one chromosome in excess of the normal somatic complement of the species; one chromosome has three homologous in place of the normal two; denoted by $2n + 1$.

Trisomic: An individual with one chromosome more than the normal chromosome complement.

Trophic level: Position in the food chain including the first level which are decomposers (bacteria, fungi); second level which are producers (plants); third level which are primary consumers (fish, insects); and fourth level which are secondary consumers (carnivores, birds).

Trough: A trough (of low pressure) is a pressure future of the synoptic chart: it is characterized by a system of isobars, which are concave towards a depression and have maximum curvature along the axis of the trough, or trough line. The trough is said to be deep or swallow, according as the maximum curvature the isobars along the trough line is more or small, respectively, the former corresponds to the V shape referred to in the obsolete term 'V-shaped' depression.

Truck crops: Crops which yield in tonnes and grown for distant market requiring heavy transport.

True allelopathy: The active substance may be released as such from some plant organ causing inhibition of other plants.

True leaf: Leaves which are not cotyledons.

Turner's syndrome: In human beings; individuals having *XO* chromosome constitution, being phenotypically female but having rudimentary sexual organs and mammary glands.

U

Umkehr effect: An effect, which is used to infer the vertical distribution of ozone from surface measurement.

Unconscious competence: The stage in the learning process where the trainees become very much competent in the learned skill and acquire unconscious competence to perform the task.

Unconscious incompetence: The stage in the learning process where the trainees are completely unaware of those learning skills closer at hands.

Univalent: An unpaired chromosome during prophase I and metaphase I of meiosis.

Upland: Cultivable land situated on high ground or the land, which has perfect natural drainage.

Upland rice: Rice grown in rainfed, naturally well drained soils, without surface water accumulation, normally without pheratic water supply and normally not bunded.

Upslope fog: Fog which is formed on the windward slopes of high ground by the forced uplift of stable, moist air till saturation is reached by adiabatic expansion.

Upstream: It is the direction opposite to the flow of the stream. The term is also used to refer to the area above the barrier place across the slope.

Urea [$CO(NH_2)_2$]: A synthetic, non-protein organic compound, crystalline or made into granules or prills for fertilizer use and containing 46 per cent nitrogen.

Urease: An universally occurring enzyme which acts on urea to produce CO_2 and NH_3.

Usufruct: The right to use and enjoy the yield of resources (land, vegetation, livestock etc.) which belong to someone else.

V

Valley: A valley is an elongated depression, sloping towards the sea or inland drainage basin. In a watershed it is usually the area drained by a river tributaries. It is the low point in a watershed that connects with the ridge to form the baseline.

Vapour: It describes a substance in the gaseous stage below its critical temperature.

Vapour drift: The movement of herbicidal vapours from the area of application to the adjacent areas.

Vapour pressure: The property which causes a chemical compound to evaporate or vapourize.

Variability: Variability refers to variability in voice, postures and use of audio-visual aids.

Variance: A measure of variability; it equals the sum of the squares of deviations of all the observations in a sample from the sample mean divided by one less than the number of observations; represented as $[\Sigma(X - X)^2]/(N - 1)$

Variation: The occurrence of differences among the individuals belonging to a single species for one or more traits.

Varieties: A variety is a population of plants having many characteristics in common. A variety may be a pureline or a mixture of purelines. Natural selection and domestication over thousands of years has resulted in the development of varieties, cultured under different seasonal, soil and climatic conditions. The long history of cultivation has thus given rise to a vast number of varieties some of them selected by cultivators themselves based on a variety of traits. Systematic efforts to improve crops began in the early 20th century with Gregor Johann Mendel. And so began the advent to modern varieties. The choice of a variety plays a very important role in ensuring a good yield, especially under rainfed conditions.

Vascular tissue: A general term referring to either or both xylem and phloem.

V-ditch: V-ditch technology is a relatively new method of land treatment which can be adopted in areas with relatively flat terrain. In this technique, ripping and angle dozing along the contours is taken up by using heavy bulldozers, which form the V-ditches. The ditches thus formed are "V" shaped; water flows easily into these drains. The V-ditch, holds water between two contour lines. Saplings are planted after leaving enough "contour interval" between the plants, so that the canopies have enough area to spread. These ditches intercept the runoff, slow it down, spread it out along the contour and let it soak into the ground, thus increasing the moisture reserves of the area as a whole. Though expensive, this method ensures around a 90 per cent chance of survival of the trees planted.

Vegetation: Sum total of plants in a given area.

Vegetative apomixis: A type of apomixis where flowers are replaced in the plant by bulbis or buds which fall to the ground like seeds.

Verification of forecasts: The process of obtaining a measure of success or forecasts by relating predicted weather to actual weather. While simple comparison may serve to reveal certain features, for example a systematic bias towards optimism or pessimism, the verification process generally consists of driving an idle by one or other of a variety of methods, which depend on the nature of the forecasts.

Vernalization: Low temperature treatment of seeds and seedlings which has the effect of artificially aging the plant and reducing the time between sowing and flowering.

Vetiver: Khus is one of the ideal plant for vegetative soil conservation measures. Commonly found in India, it is a perennial grass with extensive fibrous roots that contain an aromatic oil. It grows wild in the drier tracts of western India. In the southern states it is grown as rainfed crop in sandy loam soils. The species shows a wide range of adaptability, growing in regions of varying temperatures and soil types. Khus has gained importance as a vegetative barrier in watershed system replacing contour field bunds in some areas. It is planted in furrows at a spacing of 15 to 20 cm, along the contours. Once established, the khus contour hedges aid in the increase in field sedimentation while simultaneously reducing the erosive effect of the runoff. Its roots prevent rillying and gullying while the aromatic nature of the roots keep the rodents and other pests away.

Viability: The capability of living and developing in a given environment, *e.g.*, in the case of technology, of being practised in the long term. Also known as germination percentage. The lesser the viability of the seed, the more the seed required for an optimum population over a given area. A simple test to determine the germination percentage is to put seeds into germination pan or wet soil surface and see how many sprout in a couple of days.

Village industry: A village industry is a business enterprise on a small scale involving the skills of the local people. It provides the local population with a means of gainful employment. A village industry also serves as a means of keeping the villagers in touch with their culture and traditions. Some of the village industries include apiculture, fertilizer making crafts, leather, poultry, sericulture and pisciculture. The Khadi and Village Industries Commission and the State Khadi and Village Industries Boards are statutory bodies charged with the responsibility of promoting village industry.

Virulent phage: A virus or phage which always causes the lysis of its host bacterium following infection.

Visibility: Visibility is defined as the greatest distance at which an object of specified characteristics can be seen and identified with the unaided eye in any particular circumstances, or, in the case of night observations, could be seen and identified if the general illumination were raised to the normal daylight level.

Visual learners: Learners in whom the process is effectively related to print materials such as notes, diagrams and pictures.

Visual materials: Visual materials are instructional aids that can be seen with our eyes.

Vivipary: The phenomenon in which seed germinate in the fruit while it is still attached to the mother plant.

Volatility: It means the tendency of a chemical to vaporize or give off fumes.

Volatilization: A process where a condensed phase such as liquid or solid, is transformed into vapour by elevation of temperature or reduction of external pressure; also described as a measure of the tendency to evaporate or vaporize (from a liquid to a gas) at ordinary temperatures on exposure to the air.

W

Warm front: A front whose movement is such that the warmer air mass is replacing colder air mass.

Wasteland: These are lands, either cultivable or non-arable, that are not being put to less optimal use. Raising of trees is recommended on these lands as they will help in meeting the needs of farmers, greening the area and eventually improving the ecology and soil fertility. Large areas of our land is in some state of degradation. If this is not controlled, precious arable land will deteriorate into waste land.

Wasteland, cultivable: This refers to land that is fit for cultivation, but that which has been left fallow for a period of 5 years or more. Also known as cultivable wasteland, it differs from a "current fallow", which is left fallow only for a short period or so. The reasons for leaving the land barren are several—it could be due to an absentee landlord, personal reasons of the farmer or the lack of resources.

Wasteweirs: Wasteweirs are outlets provided in bunds to dispose of the surplus water from the fields. The objectives of wasteweirs: to provide a stable outlet for safe runoff or disposal; to prevent excessive submergence of crops on the upstream side of the bund; to prevent breaching of bunds by heavy runoff. The common type of outlet arrangement adopted at suitable places on contour bunds is clear overall, wasteweir pipe outlet, boulder or rubble wasteweir. The type of wasteweirs are located at the side bund end are channel weir, cut outlet and grass outlet.

Water balance: It is hydrological balance between rainfall and irrigation to that of changes in soil moisture evaporation, percolation and runoff.

Water conservation: Physical control, protection, management, and use of water resources in such a way as to maintain growing crop and forest lands, vegetation cover, wildlife habit and maximum sustainable benefits to people, agriculture, industry, commerce, and other segments of the national economy.

Water culture: An experimental means of determining the mineral nutrient requirement of an plant; the plant is grown with its roots dipping in solution of known composition.

Water harvesting: Collection and storage (in a tank or in soil) of rain water, either runoff or stream flow, for securing and improving water availability for crop growth under non-irrigated condition and/or animal and human consumption.

Water harvesting structures: These are storage structures for the collection of runoff from a treated or untreated land surface. The water is stored in open reservoirs or the soil itself. The objective of harvesting runoff is two fold: 1. To harvest a certain percentage of rainfall and 2. To improve the chances of survival of various rainfed production systems. There are two types of water, harvesting structures—farm ponds and percolation tanks waterlogging. It is the saturation of soil with water, causing the water table to rise high enough to expel normal soil air and interfere with plant growth or cultivation. It is a major irrigation problem associated with poor drainage. It causes soil degradation.

Water holding capacity: The weight of the water held by a given quantity of absolute dry soil when saturated.

Water killing: Relatively high water transpiration rates in evergreens during a period when absorption of water can proceed only at a relatively slow rate may lead to a type of cold injury called winter killing.

Water requirement of crops: It is that quantity of water regardless of its source, required by a crops in a given period of time for its maturity, and it includes losses due to ET plus the unavoidable losses during application of irrigation water and water required for special operations such as land preparation, puddling, leaching etc., and is expressed in depth for given time.

Water saturation deficit (WSD): Parameter for measuring water deficit in plants. It is calculated from the leaves as WSD = Fully turgid weight – Dry weight/Fully turgid weight – Dry weight.

Water yield: It is the total annual runoff volume.

Water-logging: A condition of land where the ground water soil water content reaches a level that is detrimental to plants and other useful soil micro-flora.

Watershed: A watershed is a hydrological unit or a piece of land that drains at a common point. This natural unit is evolved through the interaction or rainwater with land mass and typically comprises of arable lands, non-arable lands and natural drainage lines in rainfed areas. A watershed may range from 4-5 hectares to thousands of square kilometres. Smaller watersheds are more convenient to manage and plan for. Big ones may be justified in case they comprise mostly of grasslands and forests. In areas where agriculture is intensive, the size should be small because there are more livelihoods involved and the work that is carried out is greater. In some cases a watershed may also have within it differing topography and soil types. Based on their sizes watersheds once delineated are divided into many sub-watersheds. These are further divided into micro watersheds and mini-watersheds. Planning is usually done at the micro-watershed level. And if the micro watershed is too unwieldy to manage effectively, it may be done at the mini-watershed level.

Watershed management: The administration and regulation of the aggregate resources of a watershed for conserving water and controlling erosion and stream flow of floods is called watershed management. The objective of watershed management is to meet the problems of land and water use not in terms of anyone resource, but on the basis that all the resources are interdependent and therefore must be considered together. Watershed management calls for the following measures for integrated watershed planning—agronomic measures of soil conservation on agricultural lands; mechanical measures of soil conservation on agricultural lands; tree canopy improvements; pastureland management; gully control; soil-conservation dams; control of stream bank erosion; small dams for farm ponds and supplemental reservoirs for irrigation; irrigation works; small dams for interposing intermediate base levels of stream erosion; dams and other structures for the control of mountain torrents; works of artificial recharge of ground water; protection of wild life habitats; hydro-electric works; regulation of flood waters by dams and dyes

Watershed map: A watershed map is the combination of a cadastral map and a toposheet for that particular watershed boundary. Since the toposheets and cadastral maps are on two different scales they are brought to one scale either by enlarging the toposheet or reducing the village map. The two are superimposed to have a map of the watershed combining contours, drainage lines and other features in the toposheet, and survey numbers on the cadastral map. It is the first step in the delineation of a watershed.

Watershed-based farming system: It involves the optimum utilisation of the catchment directly through infiltration of monsoon rainfall after runoff collection and storage or after deep percolation recovery from well for the improvement and stabilisation of agriculture on the watershed.

Waterway: It is a channel into which contour/graded bunds, bench terraces, contour trenches and furrows drain. They are normally located along field boundaries of the slope as natural depressions. They are required to safely remove the excess water drained from the fields. Waterways are best when protected with grass or any other vegetal cover, as vegetation prevents the bed from getting scoured. They should also be wide and shallow.

Weather: The state of atmosphere with respect to wind, temperature, cloudiness, relative humidity, pressure etc. at a given time.

Weathering: The physical and chemical disintegration and decomposition of rocks and materials during soil genesis.

Weed: A plant in a place where it is not wanted by humans.

Weed competition index: The reduction in crop yield due to presence of weed competition in comparison to weed free plot.

Weed control: (*a*) The process of limiting weed infestations so that crops can be grown profitably or other operations can be conducted efficiently; (*b*) the process reducing weed growth or weed infestation to an acceptable level.

Weed control efficiently: It is the relative efficiency of weed control practices in respect of weed population or crop yields.

Weed eradication: The complete elemination of all live plants, plant parts, and weed seeds from an area.

Weed flora: Weed flora of a region represents the sum total of different types and kind of weeds.

Weed management: Rational development of appropriate technology to minimize the impact of weeds, provide systematic management of weed problems, and optimize intended land use.

Weed prevention: Stopping of weeds from invading and contaminating new area.

Wet bulb depression: Difference between simultaneous reading of wet and dry bulb thermometers.

Wet planting: A method of planting in which irrigation is given before planting and in usually followed in sandy soils.

Wet rice cultivation (WRC): Rainfed lowlands rice cultivation where bunds separate the fields with rice often submerged for part of the crop cycle. Apatini tribes in Subansiri district of Arunachal Pradesh who practice wet rice cultivation, which involves land and water management. Every stream rising from the surroundings hills is tapped soon after it emerges from the forest channelled at the rim of the valley and diverted by a network of primary, secondary and tertiary channel. Nutrient and fertility management of the terraces is done mainly through the recycling of agricultural waste.

Wet season: A period during which the precipitation in excess of water requirement and water accumulated in the soil and reservoirs.

Wet terrace rice cultivation (WTRC): Wet rice cultivation found on steep slopes which have been contoured to create terraces; typically much less water is available for cultivation.

Wet year: If the rainfall exceeds twice the normal deviation at a particular place, that year is said to be a wet year.

Wettable powder (Persticide): Wettable powder has an appearance similar to a dust but contains a wetting agent that disperses and suspends it when mixed with water and applied as a liquid spray.

Wetting agent: A compound which when added to a spray solution causes it to contact plant surface more thoroughly. Or, Chemicals when mixed with sprays enable to adhere more firmly to the leaves. They are also called as "spreaders" or "surface active agents" and their main function is to reduce the surface tension.

Wild type: The phenotype common in the wild population of a species or found normally in a species.

Wilting point: It is soil moisture condition at which the release of water to the plant roots is just barley too small to counterbalance the transpiration losses. On an oven-dry basis, it is the moisture content of soil at which plants wilt and fail to recover their turgidity when placed in a dark humid atmosphere. It is also called wilting coefficient or permanent wilting point.

Wind: The horizontal movement of the air related to the surface of the earth. In meteorology, the specified wind direction is that, relative to true geographic north, from which the wind blows.

Wind blast: Injury to leaves and twigs by strong winds.

Wind Break: See shelter Belt.

Wind direction: Northerly-wind from 337.6 to 22.5 degrees; North Easterly-wind from 22.6 to 67.5 degrees; Easterly-wind from 67.6 to 112.5 degrees; South Easterly-wind from 112.6 to 157.5 degrees; Southerly-wind from 157.6 to 202.5 degrees; South Westerly-wind from 202.6 to 247.5 degrees; Westerly-wind from 247.6 to 292.5 degrees; North Westerly-wind from 292.6 to 337.5 degrees.

Wind energy: It is also a form of solar energy since winds are caused by variations in the amount of heat that sun sends to different parts of the earth. Electricity is produced when a wind will catch the wind and revolves, rotating a turbine which powers electricity generator.

Wind erosion: The detachment, transportation, and deposition of soil by the action of wind. The removal and redeposition may be in more or less uniform layers or as localised below out and dines.

Windmill: It is a machine with a motor that is moved slowly by the wind to produce mechanical power or pump water. A windmill consists of a tower of about 7 mt, with a head vane, help vane, tail vane, rotor, and blades moved on tower pipe axes. Wind power is derived with use of windmills which converts wind energy into electricity. They can be located in areas where there is adequate wind pressure that can be harnessed to generate power from domestic, agricultural and industrial consumption, wind velocity of more than 7 km per hour, no high-rise structures, trees etc. around 100 mt radius of the windmill and 5-6 hrs of wind speed a day is sufficient to operate the windmill. The energy harnessed through windmills can be used for generating electricity, drawing water from underground wells upto depths of 17-18 m, minor irrigation and farm applications like winnowing, shelling and cutting. The government provides 100 per cent subsidy for installation of windmills civilworks for installation should be borne by the beneficiaries.

Wind-vane: Device used to indicate the directions from which the wind is blowing.

Winnower: A machine with one or two sieves and fan using air stream across falling grain, to separate seed and chaff.

Winnowing: The process of separation of grains from the mixture of grain and chaff or straw or *bhusa*.

Winter annual: The plant that starts germination in the fall (after the summer), lives over the winter and completes its growth, including seed production, the following season.

Wobble Hypothesis: A hypothesis to account for the degeneracy of genetic code; the first two basis of a codon and its tRNA anticodon pair properly or specifically, while the pairing at the third base has some play or wobble so that the same anticodon may pair with two or more codons differing at the third base.

Woody plants: Plants that develop woody tissues.

Workshops: A workshop refers to assembly of interested and select group of people to learn and practice specific skills under experienced professional guidance.

X

X **Chromosome:** The sex chromosome which occurs in both male and female individuals but in different numbers, *e.g.*, human male has one *X* chromosome, while females have two; in most animals and many plants, it has the genes for femaleness.

X_1, X_2: The first and second generations after X-ray treatment.

Xanthophyll: A yellow or orange carotioid pigment associated with chlorophyll in chloroplasts, also present in certain chromoplasts.

Xenia: The direct effect of pollen on the endosperm.

Xerophyte: A plant with structural and physiological features, which permit it to grow in a dry habitat, or under water stress conditions.

Xylem: The principal water-conducting non-living tissue in vascular plants characterized by the presence of tracheids; may also contain vessels, parenchyma cells, fibres, and sclereids.

Y

Y **Chromosome:** The sex chromosome which occurs either in the males (man, rodents, *Drosophila* etc.) or females (birds, reptiles) of a species, but never in both the sexes; usually the major portion of Y is genetically inactive; may or may not determine maleness depending on the species.

Yield components: Components which finally make up or control yield of any crop. For example, yield components of cereal grain yield are grain weight per grain number/ear, ear number/ plant and plant number/unit area.

Yield evaluation: Yield evaluation is the estimation of yield per hectare of principal crops. It is conducted by the Bureau of Economics and Statistics in collaboration with the state departments of agriculture. The process of yield evaluation is through a systematic survey. Yield is estimated by harvesting threshing, cleaning and weighing a measured sample area. The size of the sample area depends on the uniformity of the stand of the crop. If it is quite uniform, the field is divide into 4 quadrants. A Sample of 1 metre is then taken per quadrants. A sample of 1 × 1 metre is then taken per quadrant as also from the central section of the field. Sometimes, the harvested crop from all 5 areas of there field is mixed together, the yield is then estimated from this sample. If the stand is uneven then two samples of 1 m × 1 m are taken from each class. Two from the good patch, two from the moderate patch and two from the poor patch. The yield is then estimated from the sample.

Yield maximisation: Agronomic practices adapted to get highest possible crop production per unit area per unit time without considering either the cost of production or net return.

Yield potential: Full production capability of crop.

Yield stability: The maintenance of yield at desired level over a period of time. Its index is measured by inverse of coefficient of variation, regression between yield and environmental index and disaster level estimated by probability of a system.

Yields of crop: The harvested produce obtained from a crop grown in an unit area of land usually expressed as quintals or tonnes per hectare.

Z

Zabo: Zabo is an indigenous farming system practised in Phek district of Nagaland which has a combination of forest, agriculture, livestock and fisheries with well organised soil and water conservation base. 'Zabo' means impounding of water.

Zero tillage: The extreme in conservation tillage is no-till (zero tillage) wherein the new crop is planted in the residue of the previous crop without any prior soil tillage or seed bed preparation and is usually possible when all weeds are controlled by the use of herbicides.

Zinc deficiency: Symptoms appear 2-4 weeks after sowing as blanching of the midrib of the emerging leaf, especially at base. Brown spots appear on the older leaves which later on coahesee to give brown colour. Tillering and growth are depressed. In severe deficiency the plant dies. Zinc deficiency is associated with calcareous, alkali, peat, and volcanic soils and incidence is more severe where high rates of nitrogen and phosphorus are applied.

Zygomorphic: When a flower is divisible in equal halves from a single plane only.

Zygote: The cell derived through fusion of one male and one female gamete; it has $2n$ chromosome complement as opposed to the n for the gametes; it develops into a new individual through mitosis (higher plants and animals) or may undergo meiosis to yield the haploid phase (*e.g.*, fungi).

Zygotene: A substage of prophase I of meiosis following leptotene and before pachytene; homologous chromosomes pair during this stage to form bivalents.

